T0189967

Mathematics Manual for Water and Wastewater Treatment Plant Operators

Second Edition

Basic Mathematics for Water and Wastewater Operators

Mathematics Manual for Water and Wastewater Treatment Plant Operators, Second Edition

---

Basic Mathematics for Water and Wastewater Operators

Water Treatment Operations: Math Concepts and Calculations

Wastewater Treatment Operations: Math Concepts and Calculations

Mathematics Manual for Water and Wastewater Treatment Plant Operators

Second Edition

Basic Mathematics for Water and Wastewater Operators

Frank R. Spellman

CRC Press is an imprint of the
Taylor & Francis Group, an **informa** business

CRC Press
Taylor & Francis Group
6000 Broken Sound Parkway NW, Suite 300
Boca Raton, FL 33487-2742

Printed on acid-free paper
Version Date: 20140820

International Standard Book Number-13: 978-1-4822-2425-2 (Paperback)

Library of Congress Cataloging-in-Publication Data

Spellman, Frank R.
Mathematics manual for water and wastewater treatment plant operators / author, Frank R. Spellman. -- Second edition.
volumes cm
Includes bibliographical references and index.
Contents: volume 1. Basic mathematics for water and wastewater operators -- volume 2. Water treatment operations : math concepts and calculations -- volume 3. wastewater treatment operations : math concepts and calculations.
ISBN 978-1-4822-2421-4 (paperback)
1. Water--Purification--Mathematics. 2. Water quality management--Mathematics. 3. Water--Purification--Problems, exercises, etc. 4. Water quality management--Problems, exercises, etc. 5. Sewage--Purification--Mathematics. 6. Sewage disposal--Mathematics. 7. Sewage--Purification--Problems, exercises, etc. 8. Sewage disposal--Problems, exercises, etc. I. Title.

TD430.S64 2014
628.101'51--dc23 2013050298

Visit the Taylor & Francis Web site at
http://www.taylorandfrancis.com

and the CRC Press Web site at
http://www.crcpress.com

Contents

Preface

Hailed on its first publication as a masterly account written in an engaging, highly readable, user-friendly style, the *Mathematics Manual for Water and Wastewater Treatment Plant Operators* has been expanded and divided into three specialized texts that contain hundreds of worked examples presented in a step-by-step format; they are ideal for all levels of water treatment operators in training and practitioners studying for advanced licensure. In addition, they provide a handy desk reference and handheld guide for daily use in making operational math computations. *Basic Mathematics for Water and Wastewater Operators* covers basic math operators and operations, *Water Treatment Operations: Math Concepts and Calculations* covers computations used in water treatment, and *Wastewater Treatment Operations: Math Concepts and Calculations* covers computations commonly used in wastewater treatment plant operations.

To properly operate a waterworks or wastewater treatment plant and to pass the examination for a waterworks/wastewater operator's license, it is necessary to know how to perform certain calculations. In reality, most of the calculations that operators at the lower level of licensure need to know how to perform are not difficult, but all operators need a basic understanding of arithmetic and problem-solving techniques to be able to solve the problems they typically encounter.

The ability to learn mathematical operations can be summed up in one word: Practice! Practice! Practice! This is old-school thinking that still applies, but let's look at another word that says the same thing but differently: Redundancy! Redundancy! Redundancy! This book offers readers both practice and redundancy. That is, many sample problems and exercises are provided, and important concepts and operations are repeated, again and again.

How about waterworks/wastewater treatment plant operators at the higher levels of licensure? Do they also need to be well versed in mathematical operations? The short answer is absolutely. The long answer is that those working in water or wastewater treatment who expect to have a successful career that includes advancement to the highest levels of licensure or certification (usually prerequisites for advancement to higher management levels) must have knowledge of math at both the basic or fundamental level and advanced practical level. It is not possible to succeed in this field without the ability to perform mathematical operations.

Keep in mind that mathematics is a universal language. Mathematical symbols have the same meaning to people speaking many different languages throughout the world. The key to learning mathematics is to learn the language, symbols, definitions, and terms of mathematics that allow us to grasp the concepts necessary to solve equations.

In *Basic Mathematics for Water and Wastewater Operators*, we introduce and review concepts critical to qualified operators at the fundamental or entry level; however, this does not mean that these are the only math concepts that a competent operator must know to solve routine operation and maintenance problems. *Water*

Treatment Operations: Math Concepts and Calculations and *Wastewater Treatment Operations: Math Concepts and Calculations* both present math operations that progressively advance to higher, more practical applications of mathematical calculations—that is, the math operations that operators at the highest level of licensure would be expected to know how to perform.

At the basic level presented in this volume, we review fractions and decimals, rounding numbers, significant digits, raising numbers to powers, averages, proportions, conversion factors, flow and detention time, and the areas and volumes of different shapes. While doing this, we explain how to keep track of units of measurement (inches, feet, gallons, etc.) during the calculations and demonstrate how to solve real-life problems that require calculations. After building a strong foundation based on theoretical math concepts (the basic tools of mathematics, including fractions, decimals, percents, areas, volumes), we move on to applied math—basic math concepts applied in solving practical water/wastewater operational problems. Even though there is considerable crossover of basic math operations used by both waterworks and wastewater operators, we separate applied math problems for wastewater and water. We do this to help operators of specific unit processes unique to waterworks and wastewater operations focus on their area of specialty.

What makes *Basic Mathematics for Water and Wastewater Operators* different from other math books available? Consider the following:

- The author has worked in and around water/wastewater treatment and taught water/wastewater math for several years at the apprenticeship level and at numerous short courses for operators.
- The author has sat at the table of licensure examination preparation boards to review, edit, and write state licensure exams.
- This step-by-step training manual provides concise, practical instruction in the math skills that operators must have to pass certification tests.
- The text is user friendly; no matter the difficulty of the problem to be solved, each operation is explained in straightforward, plain English. Moreover, several hundred sample problems are presented to enhance the learning process.

The original *Mathematics Manual for Water and Wastewater Treatment Plant Operators* was highly successful and well received, but like any flagship edition of any practical manual, there is always room for improvement. Many users have provided constructive criticism, advice, and numerous suggestions. All of these inputs from actual users have been incorporated into these new texts.

The bottom line is that the material in this text is presented in manageable chunks to make learning quick and painless, and clear explanations help readers understand the material quickly; it is packed with worked examples and exercises.

To ensure correlation to modern practice and design, we present illustrative problems in terms of commonly used waterworks/wastewater treatment operations and associated parameters and cover typical math concepts for waterworks/wastewater treatment unit process operations found in today's waterworks/wastewater treatment facilities. This text is accessible to those who have little or no experience in treatment plant math operations. Readers who work through the text systematically will be surprised at how easily they can acquire an understanding of water/wastewater math concepts, adding another critical component to their professional knowledge.

A final point before beginning our discussion of math concepts: It can be said with some accuracy and certainty that, without the ability to work basic math problems typical for water/wastewater treatment, candidates for licensure will find any attempts to successfully pass licensure exams a much more difficult proposition.

Author

Frank R. Spellman, PhD, is a retired assistant professor of environmental health at Old Dominion University, Norfolk, Virginia, and the author of more than 90 books covering topics ranging from concentrated animal feeding operations (CAFOs) to all areas of environmental science and occupational health. Many of his texts are readily available online, and several have been adopted for classroom use at major universities throughout the United States, Canada, Europe, and Russia; two have been translated into Spanish for South American markets. Dr. Spellman has been cited in more than 450 publications. He serves as a professional expert witness for three law groups and as an incident/accident investigator for the U.S. Department of Justice and a northern Virginia law firm. In addition, he consults on homeland security vulnerability assessments for critical infrastructures, including water/wastewater facilities, and conducts audits for Occupational Safety and Health Administration and Environmental Protection Agency inspections throughout the country. Dr. Spellman receives frequent requests to co-author with well-recognized experts in various scientific fields; for example, he is a contributing author to the prestigious text *The Engineering Handbook*, 2nd ed. Dr. Spellman lectures on sewage treatment, water treatment, and homeland security, as well as on safety topics, throughout the country and teaches water/wastewater operator short courses at Virginia Tech in Blacksburg. He holds a BA in public administration, a BS in business management, an MBA, and both an MS and a PhD in environmental engineering.

1 Introduction

The key to learning math can be summed up in one word: Repetition! Repetition! Repetition!

Anyone who has had the opportunity to work in waterworks and/or wastewater treatment, even for a short time, quickly learns that water/wastewater treatment operations involve a large number of process control calculations. All of these calculations are based on basic math principles. In this chapter, we introduce basic mathematical terminology and definitions and calculator operations that water/wastewater operators are required to use, many of them on a daily basis.

THE 411 ON MATHEMATICS

What is mathematics? Good question. Mathematics is numbers and symbols. Math uses combinations of numbers and symbols to solve practical problems. Every day, you use numbers to count. Numbers may be considered as representing things counted. The money in your pocket or the power consumed by an electric motor is expressed in numbers. When operators make entries in daily operating logs, they enter numbers in parameter columns, indicating the operational status of various unit processes; many of these math entries are required by the plant's NPDES Permit.

Every day, we use numbers to count. Numbers can be considered as representing things counted. The money in our pocket or the power consumed by an electric motor is expressed in numbers. Again, we use numbers every day. Because we use numbers every day, we are all mathematicians—to a point.

In water/wastewater treatment, we need to take math beyond "to a point." We need to learn, understand, appreciate, and use mathematics. Probably the greatest single cause of failure to understand and appreciate mathematics is in not knowing the key definitions of the terms used. In mathematics, more than in any other subject, each word used has a definite and fixed meaning.

The following basic definitions will aid in understanding the material that follows.

Math Terminology and Definitions

- An *integer*, or an *integral number*, is a whole number; thus, 1, 2, 3, 4, 5, 6, 7, 8, 9, 10, 11, and 12 are the first 12 positive integers.
- A *factor*, or *divisor*, of a whole number is any other whole number that exactly divides it; thus, 2 and 5 are factors of 10.
- A *prime number* in math is a number that has no factors except itself and 1; examples of prime numbers are 1, 3, 5, 7, and 11.

- A *composite number* is a number that has factors other than itself and 1. Examples of composite numbers are 4, 6, 8, 9, and 12.
- A *common factor*, or *common divisor*, of two or more numbers is a factor that will exactly divide each of them. If this factor is the largest factor possible, it is called the *greatest common divisor*. Thus, 3 is a common divisor of 9 and 27, but 9 is the greatest common divisor of 9 and 27.
- A *multiple* of a given number is a number that is exactly divisible by the given number. If a number is exactly divisible by two or more other numbers, it is a common multiple of them. The least (smallest) such number is called the *lowest common multiple*. Thus, 36 and 72 are common multiples of 12, 9, and 4; however, 36 is the lowest common multiple.
- An *even number* is a number exactly divisible by 2; thus, 2, 4, 6, 8, 10, and 12 are even integers.
- An *odd number* is an integer that is not exactly divisible by 2; thus, 1,3, 5, 7, 9, and 11 are odd integers.
- A *product* is the result of multiplying two or more numbers together; thus, 25 is the product of 5 × 5. Also, 4 and 5 are factors of 20.
- A *quotient* is the result of dividing one number by another; for example, 5 is the quotient of 20 ÷ 4.
- A *dividend* is a number to be divided; a *divisor* is a number that divides; for example, in 100 ÷ 20 = 5, 100 is the dividend, 20 is the divisor, and 5 is the quotient.
- *Area* is the area of an object, measured in square units.
- *Base* is a term used to identify the bottom leg of a triangle, measured in linear units.
- *Circumference* is the distance around an object, measured in linear units. When determined for other than circles, it may be called the *perimeter* of the figure, object, or landscape.
- *Cubic units* are measurements used to express volume, cubic feet, cubic meters, etc.
- *Depth* is the vertical distance from the bottom of the tank to the top. This is normally measured in terms of liquid depth and given in terms of sidewall depth (SWD), measured in linear units.
- *Diameter* is the distance from one edge of a circle to the opposite edge passing through the center, measured in linear units.
- *Height* is the vertical distance from the base or bottom of a unit to the top or surface.
- *Linear units* are measurements used to express distances: feet, inches, meters, yards, etc.
- *Pi* (π) is a number in calculations involving circles, spheres, or cones (π = 3.14).
- *Radius* is the distance from the center of a circle to the edge, measured in linear units.
- *Sphere* is a container shaped like a ball.
- *Square units* are measurements used to express area, square feet, square meters, acres, etc.

- *Volume* is the capacity of the unit (how much it will hold), measured in cubic units (cubic feet, cubic meters) or in liquid volume units (gallons, liters, million gallons).
- *Width* is the distance from one side of the tank to the other, measured in linear units.

Calculation Steps

Standard methodology used in making mathematical calculations includes the following:

1. If appropriate, make a drawing of the information in the problem.
2. Place the given data on the drawing.
3. Ask "What is the question?" followed by "What are they really looking for?"
4. If the calculation calls for an equation, write it down.
5. Fill in the data in the equation—look to see what is missing.
6. Rearrange or transpose the equation, if necessary.
7. If available, use a calculator.
8. Always write down the answer.
9. Check any solution obtained. Does the answer make sense?

Note: Solving word math problems is difficult for many operators. Solving these problems is made easier, however, by understanding a few key words.

Key Words

- *Of* means to multiply.
- *And* means to add.
- *Per* means to divide.
- *Less than* means to subtract.

Calculators

The old saying "Use it or lose it" amply applies to mathematics. Consider a person who first learned to perform long division, multiplication, square roots, adding and subtracting, converting decimals to fractions, and other math operations using nothing more than pencil and paper and his or her own brain power. Eventually, this same person is handed a pocket calculator that can produce all of these functions and much more simply by manipulating certain keys on a keyboard. This process involves little brainpower—nothing more than punching in correct numbers and operations to achieve an almost instant answer. Backspacing to our earlier statement of "Use it or lose it" makes our point. As with other learned skills, how proficient we remain at performing a learned skill is directly proportional to the amount of time we spend using the skill—whatever that might be. We either use it or lose it. The consistent use of calculators has caused many of us to forget how to perform basic math operations with pencil and paper—for example, how to perform long division.

There can be little doubt that the proper use of a calculator can reduce the time and effort required to perform calculations; thus, it is important to recognize the calculator as a helpful tool, with the help of a well-illustrated instruction manual, of course. The manual should be large enough to read, not an inch by an inch by a quarter of an inch in size. It should have examples of problems and answers with illustrations. Careful review of the instructions and working through example problems are the best ways to learn how to use the calculator.

Keep in mind that the calculator you select should be large enough so that you can use it. Many of the modern calculators have keys so small that it is almost impossible to hit just one key. You will be doing a considerable amount of work during this study effort—make it as easy on yourself as you can.

Another significant point to keep in mind when selecting a calculator is the importance of purchasing a unit that has the functions you need. Although a calculator with a lot of functions may look impressive, it can be complicated to use. Generally, the wastewater plant operator requires a calculator that can add, subtract, multiply, and divide. A calculator with a parentheses function is helpful, and, if you must calculate geometric means for fecal coliform reporting, logarithmic capability is helpful.

In many cases, calculators can be used to perform several mathematical functions in succession. Because various calculators are designed using different operating systems, you must review the instructions carefully to determine how to make the best use of the system.

Finally, it is important to keep a couple of basic rules in mind when performing calculations:

- Always write down the calculations you wish to perform.
- Remove any parentheses or brackets by performing the calculations inside first.

DIMENSIONAL ANALYSIS

Dimensional analysis is a problem-solving method that uses the fact that any number or expression can be multiplied by 1 without changing its value. It is a useful technique used to check if a problem is set up correctly. In using dimensional analysis to check a math setup, we work with the dimensions (units of measure) only—not with numbers. To use the dimensional analysis method, we must know how to perform three basic operations. Unit factors may be made from any two terms that describe the same or equivalent amounts of what we are interested in; for example, we know that 1 inch = 2.54 centimeters.

Basic Operation 1

To complete a division of units, always ensure that all units are written in the same format; it is best to express a *horizontal fraction* (such as gal/ft^2) as a *vertical fraction*:

$$\text{gal/ft}^3 = \frac{\text{gal}}{\text{ft}^3}$$

$$\text{pounds per square inch (psi)} = \frac{\text{lb}}{\text{in.}^2}$$

The same procedures are applied in the following examples.

$$\text{ft}^3/\text{min} = \frac{\text{ft}^3}{\text{min}}$$

$$\text{sec/min} = \frac{\text{sec}}{\text{min}}$$

Basic Operation 2

We must know how to divide by a fraction. For example,

$$\frac{\left(\dfrac{\text{lb}}{\text{day}}\right)}{\left(\dfrac{\text{min}}{\text{day}}\right)} = \frac{\text{lb}}{\text{day}} \times \frac{\text{day}}{\text{min}}$$

In the above, notice that the terms in the denominator were inverted before the fractions were multiplied. This is a standard rule that must be followed when dividing fractions.

Another example is

$$\frac{\text{mm}^2}{\left(\dfrac{\text{mm}^2}{\text{m}^2}\right)} = \text{mm}^2 \times \frac{\text{m}^2}{\text{mm}^2}$$

Basic Operation 3

We must know how to cancel or divide terms in the numerator and denominator of a fraction. After fractions have been rewritten in the vertical form and division by the fraction has been reexpressed as multiplication, as shown above, then the terms can be canceled (or divided) out.

> ***Note:*** For every term that is canceled in the numerator of a fraction, a similar term must be canceled in the denominator and *vice versa*, as shown below:

$$\frac{kg}{\cancel{day}} \times \frac{\cancel{day}}{min} = \frac{kg}{min}$$

$$\cancel{mm}^2 \times \frac{m^2}{\cancel{mm}^2} = m^2$$

$$\frac{\cancel{gal}}{min} \times \frac{ft^3}{\cancel{gal}} = \frac{ft^3}{min}$$

How are units that include exponents calculated? When written with exponents, such as ft^3, a unit can be left as is or put in expanded form, (ft)(ft)(ft), depending on other units in the calculation. The point is that it is important to ensure that square and cubic terms are expressed uniformly (e.g., sq ft, ft^2, cu ft, ft^3). For dimensional analysis, the use of exponents is preferred.

For example, let's say that we wish to convert 1400 ft^3 volume to gallons, and we will use 7.48 gal/ft^3 in the conversions. The question becomes do we multiply or divide by 7.48? We can use dimensional analysis to help us answer that question.

To determine if the math setup is correct, only the dimensions are used. First, try dividing the dimensions:

$$\frac{ft^3}{gal/ft^3} = \frac{ft^3}{\left(\dfrac{gal}{ft^3}\right)}$$

Multiply the numerator and denominator to get:

$$\frac{ft^6}{gal}$$

So, by dimensional analysis, we have determined that if we divide the two dimensions (ft^3 and gal/ft^3) then the units of the answer are ft^6/gal, not gal. It is clear that division is not the right approach to making this conversion. What would have happened if we had multiplied the dimensions instead of dividing?

$$ft^3 \times gal/ft^3 = ft^3 \times \frac{gal}{ft^3}$$

Multiply the numerator and denominator to obtain:

$$\frac{ft^3 \times gal}{ft^3}$$

and cancel common terms to obtain:

$$\frac{\cancel{ft^3} \times gal}{\cancel{ft^3}} = gal$$

By multiplying the two dimensions (ft^3 and gal/ft^3), the answer will be in gallons, which is what we want. Thus, because the math setup is correct, we would then multiply the numbers to obtain the number of gallons:

$$(1400\ ft^3) \times (7.48\ gal/ft^3) = 10{,}472\ gal$$

Now, let's try another problem with exponents. We wish to obtain an answer in square feet. If we are given the two terms—70 ft^3/sec and 4.5 ft/sec—is the following math setup correct?

$$(70\ ft^3/sec) \times (4.5\ ft/sec)$$

First, only the dimensions are used to determine if the math setup is correct. By multiplying the two dimensions, we get:

$$ft^3/sec \times ft/sec = \frac{ft^3}{sec} \times \frac{ft}{sec}$$

Multiply the terms in the numerators and denominators of the fraction:

$$\frac{ft^3 \times ft}{sec \times sec} = \frac{ft^4}{sec^2}$$

The math setup is incorrect because the dimensions of the answer are not square feet; therefore, if we multiply the numbers as shown above, the answer will be wrong. Let's try division of the two dimensions instead:

$$ft^3/sec = \frac{\left(\dfrac{ft^3}{sec}\right)}{\left(\dfrac{ft}{sec}\right)}$$

Invert the denominator and multiply to get:

$$\frac{ft^3}{sec} \times \frac{sec}{ft} = \frac{\cancel{ft} \times ft \times ft \times \cancel{sec}}{\cancel{sec} \times \cancel{ft}} = ft^2$$

Because the dimensions of the answer are square feet, this math setup is correct; therefore, by dividing the numbers as was done with units, the answer will also be correct:

$$\frac{70 \text{ ft}^3/\text{sec}}{4.5 \text{ ft/sec}} = 15.56 \text{ ft}^2$$

■ Example 1.1

Problem: We are given two terms, 5 m/sec and 7 m², and the answer to be obtained should be in cubic meters per second (m³/sec). Is multiplying the two terms the correct math setup?

Solution:

$$\text{m/sec} \times \text{m}^2 = \frac{\text{m}}{\text{sec}} \times \text{m}^2$$

Multiply the numerator and denominator of the fraction:

$$\frac{\text{m} \times \text{m}^2}{\text{sec}} = \frac{\text{m}^3}{\text{sec}}$$

Because the dimensions of the answer are cubic meters per second (m³/sec), the math setup is correct; therefore, multiply the numbers to get the correct answer:

$$5 \text{ m/sec} \times 7 \text{ m}^2 = 35 \text{ m}^3/\text{sec}$$

■ Example 1.2

Problem: The flow rate in a water line is 2.3 ft³/sec. What is the flow rate expressed as gallons per minute?

Solution: Set up the math problem and then use dimensional analysis to check the math setup:

$$(2.3 \text{ ft}^3/\text{sec}) \times (7.48 \text{ gal/ft}^3) \times (60 \text{ sec/min})$$

Dimensional analysis can be used to check the math setup:

$$\text{ft}^3/\text{sec} \times \text{gal/ft}^3 \times \text{sec/min} = \frac{\cancel{\text{ft}^3}}{\cancel{\text{sec}}} \times \frac{\text{gal}}{\cancel{\text{ft}^3}} \times \frac{\cancel{\text{sec}}}{\text{min}} = \frac{\text{gal}}{\text{min}}$$

The math setup is correct as shown above; therefore, this problem can be multiplied out to get the answer in correct units:

$$(2.3 \text{ ft}^3/\text{sec}) \times (7.48 \text{ gal/ft}^3) \times (60 \text{ sec/min}) = 1032.24 \text{ gal/min}$$

2 Basic Units of Measurement and Conversions

I thought I was mathematically dysfunctional. But I am not unique—69 of out of every 9 people are also dysfunctional.

—Frank R. Spellman (2005)

$$1\ \cancel{\text{day}} \times \frac{24\ \cancel{\text{hr}}}{1\ \cancel{\text{day}}} \times \frac{60\ \text{min}}{24\ \cancel{\text{hr}}} = 1440\ \text{min}$$

SETTING THE STAGE

It is general knowledge that mathematics is the study of numbers and counting and measuring, but its associated collaterals are less well recognized. Simply, mathematics is more than numbers; it also involves the study of number patterns and relationships, and it is a way to communicate ideas. Perhaps, however, mathematics, more than anything, is a way of reasoning that is unique to human beings. No matter how we describe or define mathematics, one thing is certain: Without an understanding of mathematical units and conversion factors one might as well delve into the mysteries of deciphering hieroglyphics while blindfolded and lacking the sense of touch and reason.

Most of the calculations made in the water/wastewater industry have *units* connected or associated with them. Whereas the number tells us how many, the units tell us what we have. When we measure something, we always have to specify what units we are measuring in. For example, if someone tells us that she has 20 calcium oxide (lime), we have not learned much about what that "20" is. On the other hand, if someone tells us that she has 20 ounces, pounds, or bags of lime, then we know exactly what that person means. There are many other units of quantity we could use, such as barrels and drums. The above examples are for units of *quantity,* but there are many other things that we measure, and all of them require units.

UNITS OF MEASUREMENT

A basic knowledge of units of measurement and how to use them and convert them is essential. Water/wastewater operators should be familiar with both the U.S. customary system (USCS), or English system, and the International System of Units (SI). Some of the important units are summarized in Table 2.1, which gives some basic SI and USCS units of measurement that will be encountered.

TABLE 2.1
Commonly Used Units

Quantity	SI Units	USCS Units
Length	Meter	Foot (ft)
Mass	Kilogram	Pound (lb)
Temperature	Celsius	Fahrenheit (F)
Area	Square meter	Square foot (ft^2)
Volume	Cubic meter	Cubic foot (ft^3)
Energy	Kilojoule	British thermal unit (Btu)
Power	Watt	Btu/hr
Velocity	Meter/second	Mile/hour (mile/hr)

In the study of water/wastewater treatment plant math operations (and in actual practice), it is quite common to encounter both extremely large quantities and extremely small ones. The concentrations of some toxic substance may be measured in parts per million (ppm) or parts per billion (ppb), for example. To describe quantities that may take on such large or small values, it is useful to have a system of prefixes that accompany the units. Some of the more important prefixes are presented in Table 2.2.

Note: For comparative purposes, we like to say that 1 ppm is analogous to a full shotglass of water sitting in the bottom of a full standard-size swimming pool.

CONVERSION FACTORS

Sometimes we have to convert between different units. Suppose that a 60-inch piece of pipe is attached to an existing 6-foot piece of pipe. Joined together, how long are they? Obviously, we cannot find the answer to this question by adding 60 to 6, because the two lengths are given in different units. Before we can add the two lengths, we must convert one of them to the units of the other. Then, when we have two lengths in the same units, we can add them.

TABLE 2.2
Common Prefixes

Quantity	Prefix	Symbol
10^{-12}	Pico	p
10^{-9}	Nano	n
10^{-6}	Micro	μ
10^{-3}	Milli	m
10^{-2}	Centi	c
10^{-1}	Deci	d
10	Deca	da
10^{2}	Hecto	h
10^{3}	Kilo	k
10^{3}	Mega	M

To perform this conversion, we need a *conversion factor.* In this case, we have to know how many inches make up a foot: 12 inches. Knowing this, we can perform the calculation in two steps:

1. 60 in. is really 60 ÷ 12 = 5 ft
2. 5 ft + 6 ft = 11 ft

From the example above, it can be seen that a conversion factor changes known quantities in one unit of measure to an equivalent quantity in another unit of measure. When making the conversion from one unit to another, we must know two things:

1. The exact number that relates the two units
2. Whether to multiply or divide by that number

Confusion over whether to multiply or divide is common; on the other hand, the number that relates the two units is usually known and thus is not a problem. Understanding the proper methodology—the "mechanics"—to use for various operations requires practice and common sense.

Along with using the proper mechanics (and practice and common sense) to make conversions, probably the easiest and fastest method of converting units is to use a conversion table. The simplest conversion requires that the measurement be multiplied or divided by a constant value. For instance, if the depth of wet cement in a form is 0.85 foot, multiplying by 12 inches per foot converts the measured depth to inches (10.2 inches). Likewise, if the depth of the cement in the form is measured as 16 inches, dividing by 12 inches per foot converts the depth measurement to feet (1.33 feet).

Table 2.3 lists many of the conversion factors used in water/wastewater treatment. Note that Table 2.3 is designed with a unit of measure in the left and right columns and a constant (conversion factor) in the center column.

Note: To convert in the opposite direction (e.g., inches to feet), divide by the factor rather than multiply.

Weight, Concentration, and Flow

Using Table 2.3 to convert from one unit expression to another and *vice versa* is good practice; however, when making conversions to solve process computations in water treatment operations, for example, we must be familiar with conversion calculations based on a relationship between weight, flow or volume, and concentration. The basic relationship is

$$\text{Weight} = \text{Concentration} \times (\text{Flow or Volume}) \times \text{Factor} \tag{2.1}$$

Table 2.4 summarizes weight, volume, and concentration calculations. With practice, many of these calculations become second nature to users. The calculations are important relationships and are used often in water/wastewater treatment process control calculations, so on-the-job practice is possible.

TABLE 2.3
Conversion Table

To Convert	Multiply by	To Get
Feet	12	Inches
Yards	3	Feet
Yards	36	Inches
Inches	2.54	Centimeters
Meters	3.3	Feet
Meters	100	Centimeters
Meters	1000	Millimeters
Square yards	9	Square feet
Square feet	144	Square inches
Acres	43,560	Square feet
Cubic yards	27	Cubic feet
Cubic feet	1728	Cubic inches
Cubic feet (water)	7.48	Gallons
Cubic feet (water)	62.4	Pounds
Acre-feet	43,560	Cubic feet
Gallons (water)	8.34	Pounds
Gallons (water)	3.785	Liters
Gallons (water)	3785	Milliliters
Gallons (water)	3785	Cubic centimeters
Gallons (water)	3785	Grams
Liters	1000	Milliliters
Days	24	Hours
Days	1440	Minutes
Days	86,400	Seconds
Million gallons/day	1,000,000	Gallons/day
Million gallons/day	1.55	Cubic feet/second
Million gallons/day	3.069	Acre-feet/day
Million gallons/day	36.8	Acre-inches/day
Million gallons/day	3785	Cubic meters/day
Gallons/minute	1440	Gallons/day
Gallons/minute	63.08	Liters/minute
Pounds	454	Grams
Grams	1000	Milligrams
Pressure (psi)	2.31	Head (feet of water)
Horsepower	33,000	Foot-pounds/minute
Horsepower	0.746	Kilowatts
To Get	**Divide by**	**To Convert**

The following conversion factors are used extensively in environmental engineering (e.g., water and wastewater operations):

- 7.48 gallons = 1 cubic foot (ft^3)
- 3.785 liters = 1 gallon (gal)

TABLE 2.4
Weight, Volume, and Concentration Calculations

To Calculate	Formula
Pounds	Concentration (mg/L) × Tank volume (MG) × 8.34 lb/MG/mg/L
Pounds/day	Concentration (mg/L) × Flow (MGD) × 8.34 lb/MG/mg/L
Million gallons/day	$\frac{\text{Quantity (lb/day)}}{\text{Concentration (mg/L)} \times 8.34 \text{ lb/MG/mg/L}}$
Milligrams/liter	$\frac{\text{Quantity (lb)}}{\text{Tank volume (MG)} \times 8.34 \text{ lb/MG/mg/L}}$
Kilograms/liter	Concentration (mg/L) × Volume (MG) × 3.785 L/gal
Kilograms/day	Concentration (mg/L) × Flow (MGD) × 3.785 L/gal
Pounds/dry ton	Concentration (mg/kg) × 0.002 lb/dry ton/mg/kg

- 454 grams = 1 pound (lb)
- 1000 milliliters = 1 liter (L)
- 1000 milligrams = 1 gram (g)
- 1 ft^3/sec (cfs) = 0.6465 million gallons per day (MGD)

Note: *Density* (also called *specific weight*) is mass per unit volume and may be written as lb/ft^3, lb/gal, g/mL, or g/m^3. If we take a fixed-volume container, fill it with a fluid, and weigh it, we can determine the density of the fluid (after subtracting the weight of the container).

- 1 gallon of water weighs 8.34 pounds; the density is 8.34 lb/gal
- 1 milliliter of water weighs 1 gram; the density is 1 g/mL
- 1 cubic foot of water weighs 62.4 pounds; the density is 62.4 lb/ft^3
- 8.34 lb/gal = milligrams per liter, which is used to convert dosage in mg/L into lb/day/MGD (e.g., 1 mg/L × 10 MGD × 8.34 lb/gal = 83.4 lb/day)
- 1 psi = 2.31 feet of water (head)
- 1 foot head = 0.433 psi
- °F = 9/5(°C + 32)
- °C = 5/9(°F – 32)
- Average water usage, 100 gallons/capita/day (gpcd)
- Persons per single family residence, 3.7

CONVERSIONS

Use Tables 2.3 and 2.4 to make the conversions that are necessary in the following example problems. Other conversions are presented in appropriate sections of the chapter.

Typical Water/Wastewater Conversion Examples

■ Example 2.1

Problem: Convert 3.8 ft³/sec to gallons per second (gps).

Solution:

$$3.8\ \text{ft}^3\text{/sec} \times 7.48\ \text{ft}^3\text{/gal} = 28.42\ \text{gal/sec}$$

■ Example 2.2

Problem: Convert 2.6 ft³/sec to gallons per minute (gpm).

Solution:

$$2.6\ \text{ft}^3\text{/sec} \times 60\ \text{sec} \times 7.48\ \text{ft}^3 = 1166.9\ \text{gal/min}$$

■ Example 2.3

Problem: A treatment plant produces 7.24 million gallons per day (MGD). How many gallons per minute is that?

Solution:

$$\frac{7.24\ \cancel{\text{MG}}}{1\ \cancel{\text{day}}} \times \frac{1\ \cancel{\text{day}}}{24\ \cancel{\text{hr}}} \times \frac{1\ \cancel{\text{hr}}}{60\ \text{min}} \times \frac{1{,}000{,}000\ \text{gal}}{1\ \cancel{\text{MG}}} = 5027.78\ \text{gal/min}$$

■ Example 2.4

Problem: A pump delivers 705 gpm. How many MGD will that be?

Solution:

$$\frac{705\ \cancel{\text{gal}}}{1\ \cancel{\text{min}}} \times \frac{60\ \cancel{\text{min}}}{1\ \cancel{\text{hr}}} \times \frac{24\ \cancel{\text{hr}}}{1\ \text{day}} \times \frac{1\ \text{MG}}{1{,}000{,}000\ \cancel{\text{gal}}} = 1.0152\ \text{MGD}$$

■ Example 2.5

Problem: How many pounds of water are in a tank containing 825 gal of water?

Solution:

$$825\ \cancel{\text{gal}} \times \frac{8.34\ \text{lb}}{1\ \cancel{\text{gal}}} = 6880.5\ \text{lb}$$

■ EXAMPLE 2.6

Convert cubic feet to gallons.

$$\text{Gallons} = \text{Cubic feet (ft}^3) \times 7.48 \text{ gal/ft}^3$$

Problem: How many gallons of biosolids can be pumped to a digester that has 3600 ft³ of volume available?

Solution:

$$\text{Gallons} = 3600 \text{ ft}^3 \times 7.48 \text{ gal/ft}^3 = 26{,}928 \text{ gal}$$

■ EXAMPLE 2.7

Convert gallons to cubic feet.

$$\text{Cubic feet} = \frac{\text{Gallons}}{7.48 \text{ gal/ft}^3}$$

Problem: How many cubic feet of biosolids are removed when 18,200 gal are withdrawn?

Solution:

$$\frac{\text{Gallons}}{7.48 \text{ gal/ft}^3} = \frac{18{,}200 \text{ gal}}{7.48 \text{ gal/ft}^3} = 2433 \text{ ft}^3$$

■ EXAMPLE 2.8

Convert gallons to pounds.

$$\text{Pounds (lb)} = \text{Gallons} \times 8.34 \text{ lb/gal}$$

Problem: If 1650 gal of solids are removed from the primary settling tank, how many pounds of solids are removed?

Solution:

$$\text{Pounds} = 1650 \text{ gal} \times 8.34 \text{ lb/gal} = 13{,}761 \text{ lb}$$

■ EXAMPLE 2.9

Convert pounds to gallons.

$$\text{Gallons} = \frac{\text{Pounds}}{8.34 \text{ lb/gal}}$$

Problem: How many gallons of water are required to fill a tank that holds 7540 lb of water?

Solution:

$$\frac{\text{Pounds}}{8.34\ \text{lb/gal}} = \frac{7540\ \text{lb}}{8.34\ \text{lb/gal}} = 904\ \text{gal}$$

■ Example 2.10

Convert milligrams per liter to pounds.

> ***Note:*** For plant operations, concentrations in milligrams per liter or parts per million determined by laboratory testing must be converted to quantities of pounds, kilograms, pounds per day, or kilograms per day.

$$\text{Pounds} = \text{Concentration (mg/L)} \times \text{Volume (MG)} \times 8.34\ \text{lb/MG/mg/L}$$

Problem: The solids concentration in the aeration tank is 2580 mg/L. The aeration tank volume is 0.95 MG. How many pounds of solids are in the tank?

Solution:

$$2580\ \text{mg/L} \times 0.95\ \text{MG} \times 8.34\ \text{lb/MG/mg/L} = 20{,}441.3\ \text{lb}$$

■ Example 2.11

Convert milligrams per liter to pounds per day.

$$\text{Pounds/day} = \text{Concentration (mg/L)} \times \text{Flow (MGD)} \times 8.34\ \text{lb/MG/mg/L}$$

Problem: How many pounds of solids are discharged per day when the plant effluent flow rate is 4.75 MGD and the effluent solids concentration is 26 mg/L?

Solution:

$$26\ \text{mg/L} \times 4.75\ \text{MGD} \times 8.34\ \text{lb/mg/L/MG} = 1030\ \text{lb/day}$$

■ Example 2.12

Convert milligrams per liter to kilograms per day.

$$\text{Kilograms/day} = \text{Concentration (mg/L)} \times \text{Volume (MG)} \times 3.785\ \text{L/gal}$$

Problem: The effluent contains 26 mg/L of BOD_5. How many kilograms per day of BOD_5 are discharged when the effluent flow rate is 9.5 MGD?

Solution:

$$26\ \text{mg/L} \times 9.5\ \text{MG} \times 3.785\ \text{L/gal} = 934\ \text{kg/day}$$

■ Example 2.13

Convert pounds to milligrams per liter.

$$\text{Concentration (mg/L)} = \frac{\text{Quantity (lb)}}{\text{Volume (MG)} \times 8.34\ \text{lb/MG/mg/L}}$$

Problem: An aeration tank contains 89,990 lb of solids. The volume of the aeration tank is 4.45 MG. What is the concentration of solids in the aeration tank in milligrams per liter?

Solution:

$$\text{Concentration} = \frac{\text{Quantity (lb)}}{\text{Volume (MG)} \times 8.34\ \text{lb/MG/mg/L}}$$

$$= \frac{89{,}990\ \text{lb}}{4.45\ \text{MG} \times 8.34\ \text{lb/MG/mg/L}} = 2425\ \text{mg/L}$$

■ Example 2.14

Convert pounds per day to milligrams per liter.

$$\text{Concentration (mg/L)} = \frac{\text{Quantity (lb/day)}}{\text{Volume (MGD)} \times 8.34\ \text{lb/MG/mg/L}}$$

Problem: The disinfection process uses 4820 pounds per day of chlorine to disinfect a flow of 25.2 MGD. What is the concentration of chlorine applied to the effluent?

Solution:

$$\text{Concentration} = \frac{\text{Quantity (lb/day)}}{\text{Volume (MGD)} \times 8.34\ \text{lb/MG/mg/L}}$$

$$= \frac{4820\ \text{lb/day}}{25.2\ \text{MGD} \times 8.34\ \text{lb/MG/mg/L}} = 22.9\ \text{mg/L}$$

■ Example 2.15

Convert pounds to flow in million gallons per day.

$$\text{Flow (MGD)} = \frac{\text{Quantity (lb/day)}}{\text{Quantity (mg/L)} \times 8.34\ \text{lb/MG/mg/L}}$$

Problem: If 9640 lb of solids must be removed per day from an activated biosolids process, and the waste activated biosolids concentration is 7699 mg/L, how many million gallons per day of waste activated biosolids must be removed?

Solution:

$$\text{Flow} = \frac{\text{Quantity (lb/day)}}{\text{Quantity (mg/L)} \times 8.34 \text{ lb/MG/mg/L}}$$

$$= \frac{9640 \text{ lb/day}}{7699 \text{ mg/L} \times 8.34 \text{ lb/MG/mg/L}} = 0.15 \text{ MGD}$$

■ Example 2.16

Convert million gallons per day to gallons per minute.

$$\text{Flow (gpm)} = \frac{\text{Flow (MGD)} \times 1{,}000{,}000 \text{ gal/MG}}{1440 \text{ min/day}}$$

Problem: The current flow rate is 5.55 MGD. What is the flow rate in gallons per minute?

Solution:

$$\text{Flow} = \frac{\text{Flow (MGD)} \times 1{,}000{,}000 \text{ gal/MG}}{1440 \text{ min/day}}$$

$$= \frac{5.55 \text{ MGD} \times 1{,}000{,}000 \text{ gal/MG}}{1440 \text{ min/day}} = 3854 \text{ gpm}$$

■ Example 2.17

Convert million gallons per day to gallons per day.

$$\text{Flow (gpd)} = \text{Flow (MGD)} \times 1{,}000{,}000 \text{ gal/MG}$$

Problem: The influent meter reads 28.8 MGD. What is the current flow rate in gallons per day?

Solution:

$$\text{Flow} = 28.8 \text{ MGD} \times 1{,}000{,}000 \text{ gal/MG} = 28{,}800{,}000 \text{ gpd}$$

■ Example 2.18

Convert million gallons per day to cubic feet per second (cfs).

$$\text{Flow (cfs)} = \text{Flow (MGD)} \times 1.55 \text{ cfs/MGD}$$

Problem: The flow rate entering the grit channel is 2.89 MGD. What is the flow rate in cubic feet per second?

Solution:

$$\text{Flow} = 2.89 \text{ MGD} \times 1.55 \text{ cfs/MGD} = 4.48 \text{ cfs}$$

■ Example 2.19

Convert gallons per minute to million gallons per day.

$$\text{Flow (MGD)} = \frac{\text{Flow (gpm)} \times 1440 \text{ min/day}}{1{,}000{,}000 \text{ gal/MG}}$$

Problem: The flow meter indicates that the current flow rate is 1469 gpm. What is the flow rate in million gallons per day?

Solution:

$$\text{Flow} = \frac{\text{Flow (gpm)} \times 1440 \text{ min/day}}{1{,}000{,}000 \text{ gal/MG}}$$

$$= \frac{1469 \text{ gpm} \times 1440 \text{ min/day}}{1{,}000{,}000 \text{ gal/MG}} = 2.12 \text{ MGD}$$

■ Example 2.20

Convert gallons per day to million gallons per day.

$$\text{Flow (MGD)} = \frac{\text{Flow (gal/day)}}{1{,}000{,}000 \text{ gal/MG}}$$

Problem: The totalizing flow meter indicates that 33,444,950 gal of wastewater have entered the plant in the past 24 hr. What is the flow rate in million gallons per day?

Solution:

$$\text{Flow} = \frac{\text{Flow (gal/day)}}{1{,}000{,}000 \text{ gal/MG}} = \frac{33{,}444{,}950 \text{ gal/day}}{1{,}000{,}000 \text{ gal/MG}} = 33.44 \text{ MGD}$$

■ Example 2.21

Convert flow in cubic feet per second to million gallons per day.

$$\text{Flow (MGD)} = \frac{\text{Flow (cfs)}}{1.55 \text{ cfs/MG}}$$

Problem: The flow in a channel is determined to be 3.89 cfs. What is the flow rate in million gallons per day?

Solution:

$$\text{Flow} = \frac{\text{Flow (cfs)}}{1.55\ \text{cfs/MG}} = \frac{3.89\ \text{cfs}}{1.55\ \text{cfs/MG}} = 2.5\ \text{MGD}$$

■ **EXAMPLE 2.22**

Problem: The water in a tank weighs 675 lb. How many gallons does it hold?

Solution: Water weighs 8.34 lb/gal; therefore,

$$675\ \text{lb} \div 8.34\ \text{lb/gal} = 80.9\ \text{gal}$$

■ **EXAMPLE 2.23**

Problem: A liquid chemical weighs 62 lb/ft^3. How much does a 5-gal can of it weigh?

Solution: Solve for specific gravity, determine lb/gal, and multiply by 5:

$$\text{Specific gravity} = \frac{\text{Weight of chemical}}{\text{Weight of water}} = \frac{62\ \text{lb/ft}^3}{62.4\ \text{lb/ft}^3} = 0.99$$

$$\text{Specific gravity} = \frac{\text{Weight of chemical}}{\text{Weight of water}}$$

$$0.99 = \frac{\text{Weight of chemical}}{8.34\ \text{lb/gal}}$$

$$0.99 \times 8.34\ \text{lb/gal} = \text{Weight of chemical}$$

$$8.26\ \text{lb/gal} = \text{Weight of chemical}$$

$$8.25\ \text{lb/gal} \times 5\ \text{gal} = 41.3\ \text{lb}$$

■ **EXAMPLE 2.24**

Problem: The weight of a wooden piling with a diameter of 16 in. and a length of 16 ft is 50 lb/ft^3. If it is inserted vertically into a body of water, what vertical force is required to hold it below the water surface?

Solution: If this piling had the same weight as water, it would rest just barely submerged. Find the difference between its weight and that of the same volume of water—that is the weight required to keep it down:

$$62.4\ \text{lb/ft}^3\ \text{(water)} - 50.0\ \text{lb/ft}^3\ \text{(piling)} = 12.4\ \text{lb/ft}^3$$

$$\text{Volume of piling} = 0.785 \times (1.33\ \text{ft})^2 \times 16\ \text{ft} = 22.22\ \text{ft}^3$$

$$12.4\ \text{lb/ft}^3 \times 22.22\ \text{ft}^3 = 275.5\ \text{lb}$$

■ **Example 2.25**

Problem: A liquid chemical with a specific gravity of 1.22 is pumped at a rate of 40 gpm. How many pounds per day are being delivered by the pump?

Solution: Solve for pounds pumped per minute, then change to pounds/day.

$$8.34\text{ lb/gal} \times 1.22 = 10.2\text{ lb/gal}$$

$$40\text{ gal/min} \times 10.2\text{ lb/gal} = 408\text{ lb/min}$$

$$408\text{ lb/min} \times 1440\text{ min/day} = 587{,}520\text{ lb/day}$$

■ **Example 2.26**

Problem: A cinder block weighs 70 lb in air. When immersed in water, it weighs 40 lb. What are the volume and specific gravity of the cinder block?

Solution: The cinder block displaces 30 lb of water; solve for cubic feet of water displaced (equivalent to volume of cinder block).

$$\frac{30\text{ lb water displaced}}{62.4\text{ lb/ft}^3} = 0.48\text{ ft}^3$$

The cinder block volume is 0.48 ft^3, which weighs 70 lb; thus,

$$70\text{ lb} \div 0.48\text{ ft}^3 = 145.8\text{ lb/ft}^3 \text{ density of cinder block}$$

$$\text{Specific gravity} = \frac{\text{Density of cinder block}}{\text{Density of water}} = \frac{145.8\text{ lb/ft}^3}{62.4\text{ lb/ft}^3} = 2.34$$

Temperature Conversions

Most water/wastewater operators are familiar with the formulas used for Fahrenheit and Celsius temperature conversions:

- °C = 5/9(°F – 32)
- °F = 9/5(°C) + 32

The difficulty arises when one tries to recall these formulas from memory. Probably the easiest way to recall these important formulas is to remember these basic steps for both Fahrenheit and Celsius conversions:

1. Add 40°.
2. Multiply by the appropriate fraction (5/9 or 9/5).
3. Subtract 40°.

Obviously, the only variable in this method is the choice of 5/9 or 9/5 in the multiplication step. To make the proper choice, you must be familiar with the two scales. The freezing point of water is 32° on the Fahrenheit scale and 0° on the Celsius scale. The boiling point of water is 212° on the Fahrenheit scale and 100° on the Celsius scale.

Note: At the same temperature, higher numbers are associated with the Fahrenheit scale and lower numbers with the Celsius scale. This important relationship helps you decide whether to multiply by 5/9 or 9/5.

Now look at a few conversion problems to see how the three-step process works.

■ Example 2.27

Problem: Suppose that we wish to convert 240°F to Celsius.

Solution: Using the three-step process, we proceed as follows:

1. Add 40°

$$240° + 40° = 280°$$

2. 280° must be multiplied by either 5/9 or 9/5. Because the conversion is to the Celsius scale, we will be moving to a number *smaller* than 280. Through reason and observation, obviously, if 280 were multiplied by 9/5, the result would be almost the same as multiplying by 2, which would double 280 rather than make it smaller. If we multiply by 5/9, the result will be about the same as multiplying by 1/2, which would cut 280 in half. Because in this problem we wish to move to a smaller number, we should multiply by 5/9:

$$(5/9) \times 280° = 156°C$$

3. Now subtract 40°.

$$156°C - 40°C = 116°C$$

Therefore, 240°F = 116°C.

■ Example 2.28

Problem: Convert 22°C to Fahrenheit.

Solution:

1. Add 40°:

$$22° + 40° = 62°$$

2. Because we are converting from Celsius to Fahrenheit, we are moving from a smaller to a larger number, and 9/5 should be used in the multiplications:

$$(9/5) \times 62° = 112°$$

3. Subtract 40:

$$112° - 40° = 72°$$

Thus, 22°C = 72°F.

Obviously, knowing how to make these temperature conversion calculations is useful, but it is generally more practical to use a temperature conversion table.

If you have no problem remembering when to use 5/9 and 9/5 in the appropriate formula, then temperature conversions can simply be made using the standard formulas listed earlier:

- °C = 5/9(°F – 32)
- °F = 9/5(°C) + 32

■ EXAMPLE 2.29

Problem: Convert 56°F to °C.

Solution:

$$(5/9) \times (56°F - 32) = (5/9) \times 24 = 13.3°C$$

■ EXAMPLE 2.30

Problem: Convert 68°F to °C.

Solution:

$$(5/9) \times (68°F - 32) = (5/9) \times 36 = 19.9°C$$

■ EXAMPLE 2.31

Problem: Convert 41°F to °C.

Solution:

$$(5/9) \times (41°F - 32) = (5/9) \times 9 = 4.9°C$$

■ EXAMPLE 2.32

Problem: Convert 52°F to °C.

Solution:

$$(5/9) \times (52°F - 32) = (5/9) \times 20 = 11.1°C$$

■ EXAMPLE 2.33

Problem: Convert 102°C to °F.

Solution:

$$[(9/5) \times 102°C] + 32 = 183.6 + 32 = 215.6°F$$

■ **EXAMPLE 2.34**

Problem: Convert 33°C to °F.

Solution:

$$[(9/5) \times 33°C] + 32 = 59.4 + 32 = 91.4°F$$

■ **EXAMPLE 2.35**

Problem: Convert 81°C to °F.

Solution:

$$[(9/5) \times 81°C] + 32 = 145.8 + 32 = 177.8°F$$

■ **EXAMPLE 2.36**

Problem: Convert 65°C to °F.

Solution:

$$[(9/5) \times 65°C] + 32 = 117 + 32 = 149°F$$

■ **EXAMPLE 2.37**

Problem: Convert 63°F to °C.

Solution:

$$(5/9) \times (63°F - 32) = (5/9) \times 31 = 17.2°C$$

■ **EXAMPLE 2.38**

Problem: Convert 67°F to °C.

Solution:

$$(5/9) \times (67°F - 32) = (5/9) \times 35 = 19.4°C$$

■ **EXAMPLE 2.39**

Problem: Convert 57°C to °F.

Solution:

$$[(9/5) \times 57°C] + 32 = 102.6 + 32 = 134.6°F$$

■ **EXAMPLE 2.40**

Problem: Convert 28°F to °C.

Solution:

$$(5/9) \times (28°F - 32) = (5/9) \times -4 = -3.4°C$$

Population Equivalent or Unit Loading Factor

Although population equivalent (PE), or unit loading factor, is a parameter usually associated with wastewater treatment, keep in mind that water (a water supply for flushing, etc.) is also involved. When a wastewater characterization study is required, pertinent data are often unavailable. When this is the case, *population equivalent* or *unit per capita loading* factors are used to estimate the total waste loadings to be treated. If we know the biochemical oxygen demand (BOD) contribution of a discharger, we can determine the loading placed on the wastewater treatment system in terms of equivalent number of people. The BOD contribution of a person is normally assumed to be 0.17 lb/day BOD. To determine the population equivalent of a wastewater flow, divide the lb/day BOD content by the lb/day BOD contributed per person (e.g., 0.17 lb/day BOD).

$$\text{Population equivalent} = \frac{\text{BOD}_5 \text{ concentration (lb/day)}}{0.17 \text{ lb/day BOD per person}} \tag{2.2}$$

■ Example 2.41

Problem: A new industry wishes to connect to the city's collection system. The industrial discharge will contain an average BOD concentration of 389 mg/L, and the average daily flow will be 72,000 gpd. What is the population equivalent of the industrial discharge?

Solution: First, convert flow rate to million gallons per day:

$$\text{Flow} = \frac{72{,}000 \text{ gpd}}{1{,}000{,}000 \text{ gal/MG}} = 0.072 \text{ MGD}$$

Next, calculate the population equivalent:

$$\text{Population equivalent} = \frac{389 \text{ mg/L} \times 0.072 \text{ MGD} \times 8.34 \text{ lb/mg/L/MG}}{0.17 \text{ lb/day BOD per person}}$$

$$= 1374 \text{ people/day}$$

■ Example 2.42

Problem: An industry proposes to discharge 3455 lb of BOD to the town sewer system. What is the population equivalent of the proposed discharge?

Solution:

$$\text{Population equivalent} = \frac{3455 \text{ lb/day}}{0.17 \text{ lb/day BOD per person}}$$

$$= 20{,}324 \text{ people}$$

■ **Example 2.43**

Problem: A 0.5-MGD wastewater flow has a BOD concentration of 1600 mg/L BOD. Using an average of 0.17 lb/day BOD per person, what is the population equivalent of this wastewater flow?

Solution:

Hint: Don't forget to convert mg/L BOD to lb/day BOD, then divide by 0.17 lb/day BOD per person.

$$\text{Population equivalent} = \frac{\text{BOD concentration (lb/day)}}{\text{0.17 lb/day BOD per person}}$$

$$= \frac{\text{1600 mg/L} \times \text{0.5 MGD} \times \text{8.34 lb/gal}}{\text{0.17 lb/day BOD per person}}$$

$$= \text{39,247 people}$$

Specific Gravity and Density

Specific gravity is the ratio of the density of a substance to that of a standard material under standard conditions of temperature and pressure. The specific gravity of water is 1.0. Any substance with a density greater than that of water will have a specific gravity greater than 1.0, and any substance with a density less than that of water will have a specific gravity less than 1.0. The difference between specific gravity and density can often be confusing. To avoid such confusion and to ensure clarity, the following definitions and key points regarding specific gravity and density are provided:

- Specific gravity
  - Compares the density of a substance to a standard density
  - Does not have units for solids and liquids

Note: Specific gravity can be used to calculate the weight of a gallon of liquid chemical.

- Density
  - Is the weight per volume
  - Is expressed in lb/ft³ for solids and gases
  - Is measured in lb/gal or lb/ft³ for liquids
  - Of water varies slightly with temperature and pressure
  - Of gases changes significantly with changes in temperature and pressure
  - Of water is 62.4 lb/ft³ or 8.34 lb/gal

$$\text{Density (lb/gal)} = \text{Specific gravity of chemical} \times \text{Water (lb/gal)} \tag{2.3}$$

■ Example 2.44

Problem: The label states that the ferric chloride solution has a specific gravity of 1.58. What is the weight of 1 gal of ferric chloride solution?

Solution:

$$8.34 \text{ lb/gal} \times 1.58 = 13.2 \text{ lb/gal}$$

■ Example 2.45

Problem: If we say that the density of gasoline is 43 lb/ft^3, what is the specific gravity of gasoline?

Solution: The specific gravity of gasoline is the comparison (or ratio) of the density of gasoline to that of water:

$$\text{Specific gravity} = \frac{\text{Density of gasoline}}{\text{Density of water}} = \frac{43 \text{ lb/ft}^3}{62.4 \text{ lb/ft}^3} = 0.69$$

Note: Because the specific gravity of gasoline is less than 1.0 (lower than the specific gravity of water), it will float in water. If the specific gravity of gasoline were greater than the specific gravity of water, it would sink.

■ Example 2.46

Problem: The density of SAE 30 motor oil is 56 lb/ft^3. Find its specific gravity.

Solution:

$$\text{Specific gravity} = \frac{\text{Density of motor oil}}{\text{Density of water}} = \frac{56 \text{ lb/ft}^3}{62.4 \text{ lb/ft}^3} = 0.90$$

■ Example 2.47

Problem: A liquid chemical has a density of 11.2 lb/gal. What is its specific gravity?

Solution:

$$\text{Specific gravity} = \frac{\text{Density of liquid chemical}}{\text{Density of water}} = \frac{112 \text{ lb/gal}}{8.34 \text{ lb/gal}} = 1.34$$

■ Example 2.48

Problem: The density of chlorine gas (Cl_2) is 0.187 lb/ft^3, and the density of air is 0.075 lb/ft^3. What is the specific gravity of Cl_2?

Solution:

$$\text{Specific gravity} = \frac{\text{Density of } Cl_2}{\text{Density of air}} = \frac{0.187 \text{ lb/ft}^3}{0.075 \text{ lb/ft}^3} = 2.49$$

■ **Example 2.49**

Problem: Find the specific gravity for a rock if the density is 162 lb/ft³.

Solution:

$$\text{Specific gravity} = \frac{\text{Density of rock}}{\text{Density of water}} = \frac{162\ \text{lb/ft}^3}{62.4\ \text{lb/ft}^3} = 2.60$$

■ **Example 2.50**

Problem: Find the specific gravity of a dry chemical that has a density of 65 lb/ft³.

Solution:

$$\text{Specific gravity} = \frac{\text{Density of dry chemical}}{\text{Density of water}} = \frac{65\ \text{lb/ft}^3}{62.4\ \text{lb/ft}^3} = 1.04$$

■ **Example 2.51**

Problem: Find the specific gravity for a liquid chemical that weighs 11.07 lb/gal.

Solution:

$$\text{Specific gravity} = \frac{\text{Density of liquid chemical}}{\text{Density of water}} = \frac{11.07\ \text{lb/gal}}{8.34\ \text{lb/gal}} = 1.3$$

■ **Example 2.52**

Problem: Find the specific gravity for an acid that weighs 10.5 lb/gal.

Solution:

$$\text{Specific gravity} = \frac{\text{Density of acid}}{\text{Density of water}} = \frac{10.5\ \text{lb/gal}}{8.34\ \text{lb/gal}} = 1.3$$

■ **Example 2.53**

Problem: Find the specific gravity for a liquid chemical that weighs 12.34 lb/gal.

Solution:

$$\text{Specific gravity} = \frac{\text{Density of liquid chemical}}{\text{Density of water}} = \frac{12.34\ \text{lb/gal}}{8.34\ \text{lb/gal}} = 1.5$$

■ Example 2.54

Problem: Find the density (lb/ft^3) of a certain oil that has a specific gravity of 0.94.

Solution:

$$\text{Density} = \text{Specific gravity} \times 62.4\ \text{lb/ft}^3 = 0.94 \times 62.4\ \text{lb/ft}^3 = 58.7\ \text{lb/ft}^3$$

■ Example 2.55

Problem: Find the density (lb/gal) of a liquid chemical that has a specific gravity of 1.555.

Solution:

$$\text{Density} = \text{Specific gravity} \times 8.34\ \text{lb/gal} = 1.555 \times 8.34\ \text{lb/gal} = 12.9\ \text{lb/gal}$$

■ Example 2.56

Problem: Find the density (lb/gal) of caustic soda, which has a specific gravity of 1.530.

Solution:

$$\text{Density} = \text{Specific gravity} \times 8.34\ \text{lb/gal} = 1.530 \times 8.34\ \text{lb/gal} = 12.8\ \text{lb/gal}$$

■ Example 2.57

Problem: Find the density (lb/ft^3) of potassium permanganate, which has a specific gravity of 2.703.

Solution:

$$\text{Density} = \text{Specific gravity} \times 62.4\ \text{lb/ft}^3 = 2.703 \times 62.4\ \text{lb/ft}^3 = 168.7\ \text{lb/ft}^3$$

■ Example 2.58

Problem: A tank holds 1440 gal of a liquid chemical. The specific gravity is 0.95. How many pounds of liquid are in the tank?

Solution:

$$\text{Density} = \text{Specific gravity} \times 8.34\ \text{lb/gal} = 0.95 \times 8.34\ \text{lb/gal} = 7.92\ \text{lb/gal}$$

$$7.92\ \text{lb/gal} \times 1440\ \text{gal} = 11{,}404.8\ \text{lb}$$

■ Example 2.59

Problem: The pump rate desired is 30 gpm, and the weight of the liquid is 82.2 lb/ft^3. How many pounds of liquid can be pumped per day?

Solution:

$$82.2 \text{ lb/ft}^3 \times (1 \text{ ft}^3/7.48 \text{ gal}) = 10.99 \text{ lb/gal}$$

$$30 \text{ gal/min} \times 1440 \text{ min/day} \times 10.99 \text{ lb/gal} = 474{,}768 \text{ lb/day}$$

■ Example 2.60

Problem: A pump delivers 26 gpm. How many pounds of water does the pump deliver in 1 minute? How many pounds per minute will the pump deliver if the liquid weighs 73.7 lb/ft³?

Solution:

$$26 \text{ gal/min} \times 8.34 \text{ lb/gal} = 216.84 \text{ lb/min}$$

$$73.7 \text{ lb/ft}^3 \times 1 \text{ ft}^3/7.48 \text{ gal} = 9.853 \text{ lb/gal}$$

$$26 \text{ gal/min} \times 9.853 \text{ lb/gal} = 256.18 \text{ lb/min}$$

■ Example 2.61

Problem: A pump delivers 16 gpm. How many pounds of water does the pump deliver in 24 hours? How many pounds per day will the pump deliver if the liquid weighs 8.3 lb/gal?

Solution:

$$16 \text{ gal/min} \times 1440 \text{ min/day} \times 8.34 \text{ lb/gal} = 192{,}153.6 \text{ lb/day}$$

$$16 \text{ gal/min} \times 1440 \text{ min/day} \times 8.3 \text{ lb/gal} = 191{,}232.0 \text{ lb/day}$$

■ Example 2.62

Problem: Compare the density of chlorine gas with the density of air. Chlorine gas weighs 0.187 lb/ft³. Standard density of air is 0.075 lb/ft³.

Solution:

$$0.187 \text{ lb/ft}^3 \div 0.075 \text{ lb/ft}^3 = 2.49$$

Because chlorine gas is 2.5 times heavier than air, it will always seek the lowest space when spilled.

Flow

Flow is expressed in many different terms (English system of measurements). The most common flow terms are

- Gallons per minute (gpm)
- Cubic feet per second (cfs)
- Gallons per day (gpd)
- Million gallons per day (MGD)

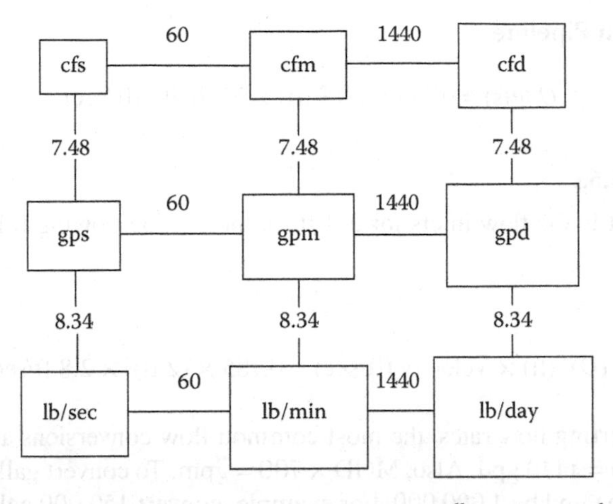

FIGURE 2.1 Flow conversions using the box method. (Adapted from Price, J.K., *Applied Math for Wastewater Plant Operators*, CRC Press, Boca Raton, FL, 1991, p. 32.)

Note: Flow is symbolized by the letter Q, and Q = Area × Velocity. Basically, flow is a volume over time.

Key Factors in Calculating Flow Problem Solutions

- Be sure the diameter (D) is squared.
- Be sure inches are converted to feet.
- Look closely at the units you are asked to find.
- Flow formulas come out in ft³/sec but you may be asked to find gpm or MGD.
- Use the flow conversion box (see Figure 2.1) or dimensional analysis to convert flows to the units desired.

Flow through a Channel

$$Q \text{ (cfs)} = \text{Width (ft)} \times \text{Depth (ft)} \times \text{Velocity (ft/sec)} \tag{2.4}$$

■ **Example 2.63**

Problem: What is the flow in cubic feet per second (cfs) for a channel that is 2 ft wide and 3 ft deep with water moving at 1.2 ft/sec?

Solution:

$$Q = \text{Width (ft)} \times \text{Depth (ft)} \times \text{Velocity (ft/sec)} = 2 \text{ ft} \times 3 \text{ ft} \times 1.2 \text{ ft/sec} = 7.2 \text{ cfs}$$

Flow through a Pipeline

$$Q \text{ (cfs)} = 0.785 \times (D)^2 \text{ (ft)} \times \text{Velocity (ft/sec)} \tag{2.5}$$

■ Example 2.64

Problem: What is the flow in cfs for a 2-ft-diameter pipe flowing full at a velocity of 2.8 ft/sec?

Solution:

$$Q = 0.785 \times (D)^2 \text{ (ft)} \times \text{Velocity (ft/sec)} = 0.785 \times (2 \text{ ft})^2 \times 2.8 \text{ ft/sec} = 8.79 \text{ cfs}$$

When converting flow rates, the most common flow conversions are 1 cfs = 449 gpm and 1 gpm = 1440 gpd. Also, MGD × 700 = gpm. To convert gallons per day to MGD, divide the gpd by 1,000,000. For example, convert 150,000 gallons to MGD:

$$150{,}000 \text{ gpd} \div 1{,}000{,}000 = 0.150 \text{ MGD}$$

In some instances, flow is given in MGD but is needed in gpm. To make the conversion from MGD to gpm, two steps are required.

1. Convert the gpd by multiplying by 1,000,000.
2. Convert to gpm by dividing by the number of minutes in a day (1440 min/day).

■ Example 2.65

Problem: Convert 0.135 MGD to gpm.

Solution:

1. Convert the flow in MGD to gpd:

$$0.135 \text{ MGD} \times 1{,}000{,}000 = 135{,}000 \text{ gpd}$$

2. Convert to gpm by dividing by the number of minutes in a day (24 hr/day × 60 min/hr = 1440 min/day):

$$135{,}000 \text{ gpd} \div 1440 \text{ min/day} = 94 \text{ gpm}$$

To determine flow through a pipeline, channel, or stream, we use the following equation:

$$Q = V \times A \tag{2.6}$$

where

Q = Cubic feet per second (cfs).
V = Velocity in feet per second (ft/sec).
A = Area in square feet (ft²).

■ **Example 2.66**

Problem: Find the flow in cubic feet per second in an 8-in. line if the velocity is 3 ft/sec.

Solution:

1. Determine the cross-sectional area of the line in square feet. Start by converting the diameter of the pipe to inches.
2. The diameter is 8 in.; therefore, the radius is 4 in. = 4/12 of a foot, or 0.33 ft.
3. Find the area (A) in square feet:

$$A = \pi \times r^2$$

$$A = \pi \times (0.33 \text{ ft})^2 = 3.14 \times 0.109 \text{ ft}^2 = 0.342 \text{ ft}^2$$

4. $Q = V \times A$:

$$Q = 3 \text{ ft/sec} \times 0.342 \text{ ft}^2 = 1.03 \text{ cfs}$$

■ **Example 2.67**

Problem: Find the flow in gallons per minute when the total flow for the day is 75,000 gpd.

Solution:

$$75{,}500 \text{ gpd} \div 1440 \text{ min/day} = 52 \text{ gpm}$$

■ **Example 2.68**

Problem: Find the flow in gpm when the flow is 0.45 cfs.

Solution:

$$0.45 \text{ cfs} \times 449 \text{ gpm/cfs} = 202 \text{ gpm}$$

■ **Example 2.69**

Problem: A system uses 3 MGD. How many gallons per minute does it use?

Solution: Convert MGD to gpm.

$$3 \text{ MGD} \times 700 = 2100 \text{ gpm}$$

■ **Example 2.70**

Problem: A pipeline has a carrying capacity of 2 cfs. How many gallons per minute can it handle?

Solution: Convert cfs to gpm.

$$2 \text{ cfs} \times 449 \text{ gpm/cfs} = 898 \text{ gpm}$$

■ **EXAMPLE 2.71**

Problem: A well pumps 410 gpm. How many million gallons per day will it pump?

Solution: Convert gpm to MGD.

$$410 \text{ gpm} \div 700 = 0.59 \text{ MGD}$$

■ **EXAMPLE 2.72**

Problem: A channel 30 in. wide has water flowing to a depth of 1.5 ft. If the velocity of the water is 2.5 ft/sec, what is the flow in the channel in ft³/sec? What is the flow in gpm?

Solution:

$$30 \text{ in.} \div 12 \text{ in./ft} = 2.5 \text{ ft}$$

$$Q = \text{Width} \times \text{Depth} \times \text{Velocity} = 2.5 \text{ ft} \times 1.5 \text{ ft} \times 2.5 \text{ ft/sec} = 9.38 \text{ ft}^3\text{/sec}$$

$$9.38 \text{ ft}^3\text{/sec} \times 60 \text{ sec/min} \times 7.48 \text{ gal/ft}^3 = 4210 \text{ gpm}$$

■ **EXAMPLE 2.73**

Problem: The flow through a 24-in. pipe is moving at a velocity of 5.5 ft/sec. What is the flow rate in gallons per minute?

Solution:

$$24 \text{ in.} \div 12 \text{ in./ft} = 2 \text{ ft}$$

$$Q = 0.785 \times (D)^2 \text{ (ft)} \times \text{Velocity (ft/sec)}$$

$$Q = 0.785 \times (2 \text{ ft})^2 \times 5.5 \text{ ft/sec} = 17.27 \text{ ft}^3\text{/sec}$$

$$17.27 \text{ ft}^3\text{/sec} \times 60 \text{ sec/min} \times 7.48 \text{ gal/ft}^3 = 7750.78 \text{ gpm}$$

■ **EXAMPLE 2.74**

Problem: A channel is 48 in. wide and has water flowing to a depth of 2 ft. If the velocity of the water is 2.6 ft/sec, what is the flow in the channel in cubic feet per second?

Solution:

$$48 \text{ in.} \div 12 \text{ in./ft} = 4 \text{ ft}$$

$$Q = 4 \text{ ft} \times 2 \text{ ft} \times 2.6 \text{ ft/sec} = 20.8 \text{ ft}^3\text{/sec}$$

■ **EXAMPLE 2.75**

Problem: A channel 3 ft wide has water flowing to a depth of 2.7 ft. If the velocity through the channel is 122 ft/min, what is the flow rate in cubic feet per minute? In million gallons per day?

Solution:

$$Q = 3 \text{ ft} \times 2.7 \text{ ft} \times 122 \text{ ft/sec} = 988.2 \text{ ft}^3\text{/sec}$$

Using the flow chart in Figure 2.1 to convert:

$$Q = 10.64 \text{ MGD}$$

■ Example 2.76

Problem: A channel is 3 ft wide and has water flowing at a velocity of 1.3 ft/sec. If the flow through the channel is 8.0 ft³/sec, what is the depth of the water in the channel in feet?

Solution:

$$8.0 \text{ ft}^3\text{/sec} = 3 \text{ ft} \times \text{Depth} \times 1.3 \text{ ft/sec}$$

$$\frac{8.0 \text{ ft}^3\text{/sec}}{3 \text{ ft} \times 1.3 \text{ ft/sec}} = \text{Depth}$$

$$2.1 \text{ ft} = \text{Depth}$$

■ Example 2.77

Problem: The flow through a 2-ft-diameter pipeline is moving at a velocity of 3.3 ft/sec. What is the flow rate in cubic feet per second?

Solution:

$$Q = 0.785 \times (2 \text{ ft})^2 \times 3.3 \text{ ft/sec} = 0.785 \times 4 \text{ ft}^2 \times 3.3 \text{ ft/sec} = 10.36 \text{ ft}^3\text{/sec}$$

■ Example 2.78

Problem: The flow through a pipe is 0.6 ft³/sec. If the velocity of the flow is 3.5 ft/sec, and the pipe is flowing full, what is the diameter of the pipe in inches?

Solution:

$$0.6 \text{ ft}^3\text{/sec} = 0.785 \times D^2 \times 3.5 \text{ ft/sec}$$

$$\frac{0.6 \text{ ft}^3\text{/sec}}{0.785 \times 3.5 \text{ ft/sec}} = D^2$$

$$0.2181 \text{ ft} = D^2$$

$$\sqrt{0.2181 \text{ ft}} = D$$

$$0.47 \text{ ft} = D$$

$$0.47 \text{ ft} \times 12 \text{ in./ft} = 5.64 \text{ in.}$$

■ Example 2.79

Problem: An 8-in.-diameter pipeline has water flowing at a velocity of 3.5 ft/sec. What is the flow rate in gpm?

Solution:

$$8 \text{ in.} \div 12 \text{ in./ft} = 0.667 \text{ ft}$$

$$Q = 0.785 \times (0.667 \text{ ft})^2 \times 3.5 \text{ ft/sec} = 1.222 \text{ ft}^3\text{/sec}$$

Use the flow chart in Figure 2.1 to convert:

$$Q = 548.4 \text{ gpm}$$

■ Example 2.80

Problem: A channel is 3 ft wide with water flowing to a depth of 3 ft. If the velocity in the channel is found to be 1.9 fps, what is the flow rate in the channel in cubic feet per second?

Solution:

$$Q = 3 \text{ ft} \times 3 \text{ ft} \times 1.9 \text{ ft/sec} = 17.1 \text{ ft}^3\text{/sec}$$

■ Example 2.81

Problem: A 12-in.-diameter pipe is flowing full. What is the ft³/min flow rate in the pipe if the velocity is 120 ft/min?

Solution:

$$12 \text{ in.} = 1 \text{ ft}$$

$$Q = 0.785 \times (1 \text{ ft})^2 \times 120 \text{ ft/min} = 94.2 \text{ ft}^3\text{/min}$$

■ Example 2.82

Problem: A water main with a diameter of 18 in. is determined to have a velocity of 184 ft/min. What is the flow rate in gallons per minute?

Solution:

$$18 \text{ in.} \div 12 \text{ in./ft} = 1.5 \text{ ft}$$

$$Q = 0.785 \times (1.5 \text{ ft})^2 \times 184 \text{ ft/min} = 324.99 \text{ ft}^3\text{/min}$$

$$324.99 \text{ ft}^3\text{/min} \times 7.48 \text{ gal/ft}^3 = 2430.93 \text{ gal/min}$$

■ Example 2.83

Problem: A 24-in. main has a velocity of 210 ft/min. What is the gpd flow rate for the pipe?

Solution:

$$24 \text{ in.} \div 12 \text{ in./ft} = 2 \text{ ft}$$

$$Q = 0.785 \times (2 \text{ ft})^2 \times 210 \text{ ft/min} = 659.4 \text{ ft}^3\text{/min}$$

$$659.4 \text{ ft}^3\text{/min} \times 1440 \text{ min/day} \times 7.48 \text{ gal/ft}^3 = 7{,}102{,}529.2 \text{ gpd}$$

■ Example 2.84

Problem: What would be the gpd flow rate for a 6-in. line flowing at 2.2 ft/sec?

Solution:

$$6 \text{ in.} \div 12 \text{ in./ft} = 0.5 \text{ ft}$$

$$Q = 0.785 \times (0.5 \text{ ft})^2 \times 2.2 \text{ ft/sec} = 0.4318 \text{ ft}^3\text{/sec}$$

$$0.4318 \text{ ft}^3\text{/sec} \times 60 \text{ sec/min} \times 1440 \text{ min/day} \times 7.48 \text{ gal/ft}^3 = 279{,}060.24 \text{ gpd}$$

■ Example 2.85

Problem: A 36-in. water main has just been installed. If the main is flushed at 2.8 ft/sec, how many gallons per minute of water should be flushed from the hydrant?

Solution:

$$36 \text{ in.} \div 12 \text{ in./ft} = 3 \text{ ft}$$

$$Q = 0.785 \times (3 \text{ ft})^2 \times 2.8 \text{ ft/sec} = 19.782 \text{ ft}^3\text{/sec}$$

$$19.782 \text{ ft}^3\text{/sec} \times 60 \text{ sec/min} \times 7.78 \text{ gal/ft}^3 = 9234.24 \text{ gpm}$$

■ Example 2.86

Problem: A 36-in. water main has just been installed. If the main is flowing at a velocity of 2.2 ft/sec, how many million gallons per day will the pipe deliver?

$$36 \text{ in.} \div 12 \text{ in./ft} = 3 \text{ ft}$$

$$Q = 0.785 \times (3 \text{ ft})^2 \times 2.2 \text{ ft/sec} = 15.54 \text{ ft}^3\text{/sec}$$

$$15.54 \text{ ft}^3\text{/sec} \times 60 \text{ sec/min} \times 1440 \text{ min/day} \times 7.48 \text{ gal/ft}^3 \times 1 \text{ MG/1,000,000 gal} = 10.04 \text{ MGD}$$

■ Example 2.87

Problem: A pipe has a diameter of 18 in. If the pipe is flowing full, and the water is known to flow a distance of 850 yards in 5 minutes, what is the MGD flow rate for the pipe?

Solution:

$$18 \text{ in.} \div 12 \text{ in./ft} = 1.5 \text{ ft}$$

$$850 \text{ yd} \times 3 \text{ ft/yd} = 2550 \text{ ft}$$

$$\text{Velocity} = 2550 \text{ ft/5 min} = 510 \text{ ft/min}$$

$$Q = 0.785 \times (1.5 \text{ ft})^2 \times 510 \text{ ft/min} = 900.7875 \text{ ft}^3\text{/min}$$

$$900.7875 \text{ ft}^3\text{/min} \times 1440 \text{ min/day} \times 7.48 \text{ gal/ft}^3 \times 1 \text{ MG/1,000,000 gal} = 9.70 \text{ MGD}$$

■ Example 2.88

Problem: A water crew is flushing hydrants on a 12-in.-diameter main. The pitot-tube gauge indicates that 580 gpm are being flushed from the hydrant. What is the flushing velocity (in ft/min) through the pipe?

Solution:

$$12 \text{ in.} \div 12 \text{ in./ft} = 1 \text{ ft}$$

$$A = 0.785 \times (1)^2 = 0.785 \text{ ft}^2$$

$$Q = 580 \text{ gpm} \times 7.48 \text{ gal/ft}^3 = 4338.4 \text{ ft}^3\text{/min}$$

$$4338.4 \text{ ft}^3\text{/min} = 0.785 \text{ ft}^2\text{/min} \times \text{Velocity}$$

$$\frac{4338.4 \text{ ft}^3\text{/min}}{0.785 \text{ ft}^2\text{/min}} = \text{Velocity}$$

$$5226.62 \text{ ft/min} = \text{Velocity}$$

■ Example 2.89

Problem: Geologic studies show that water in an aquifer moves 22 ft in 55 days. What is the average velocity of the water in feet per day?

Solution:

$$\text{Velocity} = 22 \text{ ft} \div 55 \text{ days} = 0.4 \text{ ft/day}$$

■ Example 2.90

Problem: If the water in a water table aquifer moves 2.5 ft/day, how far will the water travel in 17 days?

Solution:

$$2.5 \text{ ft/day} \times 17 \text{ days} = 42.5 \text{ ft}$$

■ **EXAMPLE 2.91**

Problem: If the water in a water table aquifer moves 2.5 ft/day, how long will it take the water to move 61 ft?

Solution:

$$61 \text{ ft} \times (1 \text{ day}/2.5 \text{ ft}) = 24.4 \text{ days}$$

■ **EXAMPLE 2.92**

Problem: The average velocity in a full-flowing pipe is measured and known to be 2.8 fps. The pipe is a 24-in. main. Assuming that the pipe flows 20 hr/day and that the month in question contains 30 days, what is the total flow for the pipe in million gallons for that one month?

Solution:

$$24 \text{ in.} \div 12 \text{ in./ft} = 2 \text{ ft}$$

$$Q = 0.785 \times (2 \text{ ft})^2 \times 2.8 \text{ ft/sec} = 8.792 \text{ ft}^3\text{/sec}$$

$$8.792 \text{ ft}^3\text{/sec} \times 60 \text{ sec/min} \times 60 \text{ min/hr} \times 20 \text{ hr/day} \times 30 \text{ days/month} \times 7.48 \text{ gal/ft}^3 = 142.05 \text{ MG}$$

DETENTION TIME

Detention time is the length of time in minutes or hours that 1 gallon of water is retained in a vessel or basin, or the period from the time the water enters a settling basin until it flows out the other end. When calculating unit process detention times, we are calculating the length of time it takes the water to flow through that unit process. The detention time formula can also be used to calculate how long it will take to fill a tank. Detention times are normally calculated for the following basins or tanks:

- Flash mix chambers (seconds)
- Flocculation basins (minutes)
- Sedimentation tanks or clarifiers (hours)
- Wastewater ponds (days
- Oxidation ditches (hours)

To calculate the detention period of a basin, the volume of the basin must first be obtained. Using a basin 25 ft wide, 70 ft long, and 12 ft deep, the volume in gallons would be

$$\text{Volume} = \text{Length} \times \text{Width} \times \text{Depth} = 70 \text{ ft} \times 25 \text{ ft} \times 12 \text{ ft} = 21{,}000 \text{ ft}^3$$

$$\text{Gallons} = \text{Volume} \times 7.48 \text{ gal/ft}^2 = 21{,}000 \text{ ft}^3 \times 7.48 \text{ gal/ft}^3 = 157{,}080 \text{ gal}$$

If we assume that the plant filters 300 gpm, then we have 157,080 gal ÷ 300 gpm = 523 minutes, or roughly 9 hours of detention time. Stated another way, the detention time is the length of time theoretically required for the coagulated water to flow through the basin. If chlorine is added to the water as it enters the basin, the chlorine contact time would be 9 hours. To determine the CT value (concentration of free chlorine residual × disinfectant contact time in minutes) used to determine the effectiveness of chlorine, we must calculate detention time.

Note: If detention time is desired in minutes, then the flow rate used in the calculation should have the same time frame (cfm or gpm, depending on whether tank volume is expressed as cubic feet or gallons). If a detention time is desired in hours, then the flow rate used in the calculation should be ft^3/hr or gal/hr.

Note: True detention time is the T portion of the CT value.

Detention time is calculated in units of time. The most common are seconds, minutes, hours, and days. Examples of detention time equations where time and volume units match include the following:

$$\text{Detention time (sec)} = \frac{\text{Volume of tank (ft}^3\text{)}}{\text{Flow rate (cfs)}} \tag{2.7}$$

$$\text{Detention time (min)} = \frac{\text{Volume of tank (gal)}}{\text{Flow rate (gpm)}} \tag{2.8}$$

$$\text{Detention time (hr)} = \frac{\text{Volume of tank (gal)}}{\text{Flow rate (gph)}} \tag{2.9}$$

$$\text{Detention time (days)} = \frac{\text{Volume of tank (gal)}}{\text{Flow rate (gpd)}} \tag{2.10}$$

The simplest way to calculate detention time is to divide the volume of the container by the flow rate into the container. The theoretical detention time of a container is the same as the amount of time it would take to fill the container if it were empty. For volume, the most common units used are gallons, although occasionally cubic feet may also be used. Time units will be in whatever the units are used to express the flow; for example, if the flow is in gpm, the detention time will be in days. If, in the final result, the detention time is in the wrong time units, simply convert to the appropriate units.

■ Example 2.93

Problem: The reservoir for the community holds 110,000 gal. The well will produce 60 gpm. What is the detention time in the reservoir in hours?

Solution:

$$\text{Detention time} = \frac{110{,}000 \text{ gal}}{60 \text{ gpm}} = 1833 \text{ min}; \quad \frac{1833 \text{ min}}{60 \text{ min/hr}} = 30.6 \text{ hr}$$

■ **EXAMPLE 2.94**

Problem: Find the detention time in a 55,000-gal reservoir if the flow rate is 75 gpm.

Solution:

$$\text{Detention time} = \frac{55{,}000\text{ gal}}{75\text{ gpm}} = 733\text{ min};\quad \frac{733\text{ min}}{60\text{ min/hr}} = 12.2\text{ hr}$$

■ **EXAMPLE 2.95**

Problem: If the fuel consumption of a boiler is 30 gpd, how many days will the 1000-gal tank last?

Solution:

$$1000\text{ gal} \div 30\text{ gpd} = 33.3\text{ days}$$

■ **EXAMPLE 2.96**

Problem: A 60,000 gal tank receives 250,000 gpd flow. What is the detention in hours?

Solution:

$$\text{Detention time} = \frac{60{,}000\text{ gal}}{250{,}000\text{ gpd}} = 0.24\text{ days};\quad \frac{0.24\text{ days}}{24\text{ hr/day}} = 5.8\text{ hr}$$

■ **EXAMPLE 2.97**

Problem: A tank is 70 ft by 60 ft by 10 ft, and the flow is 3.0 MGD. What is the detention time in hours?

Solution: Find the volume in cubic feet.

$$70\text{ ft} \times 60\text{ ft} \times 10\text{ ft} = 42{,}000\text{ ft}^3$$

Change cubic feet to gallons.

$$42{,}000\text{ ft}^3 \times 7.48\text{ gal/ft}^3 = 314{,}160\text{ gal}$$

Change MGD to gpd.

$$3\text{ MGD} = 3{,}000{,}000\text{ gpd}$$

Find the detention time in days and change days to hours.

$$\text{Detention time} = \frac{314{,}160\text{ gal}}{3{,}000{,}000\text{ gpd}} = 0.10\text{ days};\quad \frac{0.10\text{ days}}{24\text{ hr/day}} = 2.4\text{ hr}$$

■ **EXAMPLE 2.98**

Problem: A tank 100 ft in diameter and 23 ft deep. If the flow into the tank is 1500 gpm and the flow of the tank is 325 gpm, how many hours will it take to fill the tank?

Solution: Calculate the volume in cubic feet.

$$\text{Volume} = 3.14 \times (50\text{ ft})^2 \times 23\text{ ft} = 180{,}600\text{ ft}^3$$

or

$$\text{Volume} = 0.785 \times (100\text{ ft})^2 \times 23\text{ ft} = 180{,}600\text{ ft}^3$$

Change cubic feet to gallons.

$$180{,}600\text{ ft}^3 \times 7.48\text{ gal/ft}^3 = 1{,}350{,}900\text{ gal}$$

Calculate the net inflow.

$$1500\text{ gpm} - 325\text{ gpm} = 1175\text{ gpm}$$

Calculate how long until full, or detention time, and change minutes to hours.

$$\text{Detention time (min)} = \frac{1{,}350{,}900\text{ gal}}{1175\text{ gpm}} = 1150\text{ min}$$

$$1150\text{ min} \div 60\text{ min/hr} = 19.2\text{ hr}$$

CHEMICAL ADDITION CONVERSIONS

One of the most important water/wastewater operator functions is to make various chemical additions to unit processes. In this section, we demonstrate how to calculate the required amount of chemical (active ingredient and inactive ingredient), dry chemical feed rate, and liquid chemical feed rate.

Required Amount of Chemical (Active Ingredient)

$$\text{Chemical (lb/day)} = \text{Required dose (mg/L)} \times \text{Flow (MGD)} \times 8.34\text{ lb/MG/mg/L} \quad (2.11)$$

■ **EXAMPLE 2.99**

Problem: A laboratory jar test indicates a required dose of 4.1 mg/L of ferric chloride. The flow rate is 5.15 MGD. How many pounds of ferric chloride will be needed each day?

Solution:

$$\text{Chemical} = 4.1\text{ mg/L} \times 5.15\text{ MGD} \times 8.34\text{ lb/MG/mg/L} = 176.1\text{ lb/day}$$

Required Amount of Chemical (Inactive Ingredient)

Because industrial strength chemicals are normally less than 100% active ingredient, the amount of chemical must be adjusted to compensate for the inactive components.

$$\text{Required amount (lb/day)} = \frac{\text{Active ingredient required (lb/day)}}{\text{\% Active ingredient}} \tag{2.12}$$

■ Example 2.100

Problem: To achieve the desired phosphorus removal, 180 lb of ferric chloride must be added to the daily flow. The feed solution is 66% ferric chloride. How many pounds of feed solution will be needed?

Solution:

$$\text{Required amount} = \frac{\text{Active ingredient required (lb/day)}}{\text{\% Active ingredient}} = \frac{180\text{ lb/day}}{0.66} = 273\text{ lb/day}$$

Dry Chemical Feed Rate

When chemical is to be added in dry (powder, granular, etc.) form, the chemical feed rate can be expressed in units such as pounds per hour or grams per minute.

$$\text{Feed rate (lb/hr)} = \frac{\text{Required amount (lb/day)}}{24\text{ hr/day}} \tag{2.13}$$

$$\text{Feed rate (g/min)} = \frac{\text{Required amount (lb/day)} \times 454\text{ g/lb}}{1440\text{ min/day}} \tag{2.14}$$

■ Example 2.101

Problem: A plant must feed 255 lb/day of high-test hypochlorite (HTH) powder to reduce odors. What is the required feed rate in pounds per hour? In grams per minute?

Solution:

$$\text{Feed rate (lb/hr)} = \frac{255\text{ lb/day}}{24\text{ hr/day}} = 10.6\text{ lb/hr}$$

$$\text{Feed rate (g/min)} = \frac{255\text{ lb/day} \times 454\text{ g/lb}}{1440\text{ min/day}} = 80\text{ g/min}$$

Liquid Chemical Feed Rate

If chemical is fed in liquid form, the required amount (pounds, grams, etc.) of process chemical must be converted to an equivalent volume (gallons, milliliters, etc.). This volume is then converted to the measurement system of the solution feeder (gallons/day, milliliters/minute, etc.).

Note: The weight of a gallon of the process chemical is usually printed on the container label or the material safety data sheet (MSDS), or weight per gallon can be determined if the specific gravity of the chemical is supplied.

$$\text{Feed rate (gpd)} = \frac{\text{Required amount (lb/day)}}{\text{Weight (lb/gal)}} \tag{2.15}$$

$$\text{Feed rate (gpm)} = \frac{\text{Required amount (lb/day)}}{\text{Weight (lb/gal)} \times 1440 \text{ min/day}} \tag{2.16}$$

$$\text{Feed rate (mL/min)} = \frac{\text{Required amount (lb/day)} \times 3785 \text{ mL/gal}}{\text{Weight (lb/gal)} \times 1440 \text{ min/day}} \tag{2.17}$$

■ **EXAMPLE 2.102**

Problem: To achieve phosphorus removal, the plant must add 812 lb of ferric chloride feed solution each day. The ferric chloride solution weighs 11.1 lb/gal. What is the required feed rate in gallons per day? In gallons per minute? In milliliters per minute?

Solution:

$$\text{Feed rate (gpd)} = \frac{812 \text{ lb/day}}{11.1 \text{ lb/gal}} = 73 \text{ gpd}$$

$$\text{Feed rate (gpm)} = \frac{812 \text{ lb/day}}{11.1 \text{ lb/gal} \times 1440 \text{ min/day}} = 0.05 \text{ gpm}$$

$$\text{Feed rate (mL/min)} = \frac{812 \text{ lb/day} \times 3785 \text{ mL/gal}}{11.1 \text{ lb/gal} \times 1440 \text{ min/day}} = 192 \text{ mL/min}$$

WIRE-TO-WATER CALCULATIONS

In the past, wire-to-water calculations for pumping systems were seldom considered except for very large pumps where the wire-to-water efficiency of the pump–motor combination was of significant importance. The emergence of the variable speed drive, digital electronics, and higher electrical costs has made wire-to-water efficiency invaluable for smaller pumping systems. Thus, wire-to-water calculations are important in water/wastewater operations where large and small pumps are used for numerous applications (Rishel, 2001). The term "wire-to-water" refers to the conversion of electrical horsepower to water horsepower. The motor takes electrical energy and converts it into mechanical energy. The pump turns mechanical energy into hydraulic energy. The electrical energy is measured as motor horsepower (mhp). The mechanical energy is measured as brake horsepower (bhp), and the hydraulic energy is measured as water horsepower (whp).

Horsepower and Energy Costs

In water/wastewater treatment and ancillaries, horsepower is a common expression for power. One horsepower is equal to 33,000 ft-lb of work per minute. This value is determined, for example, when selecting a pump or combination of pumps to ensure adequate pumping capacity (a major use of calculating horsepower in water/wastewater treatment). Pumping capacity depends on the flow rate desired and the feet of head against which the pump must pump (effective height).

Calculations of horsepower are made in conjunction with a variety of treatment plant operations. The basic concept from which the horsepower calculation is derived is the concept of work. *Work* involves the operation of a force (lb) over a specific distance (ft). The amount of work accomplished is measured in foot-pounds (ft-lb):

$$\text{ft} \times \text{lb} = \text{ft-lb} \tag{2.18}$$

The *rate of doing work* (power) involves a time factor. Originally, the rate of doing work or power compared the power of a horse to that of a steam engine. The rate at which a horse could work was determined to be about 550 ft-lb/sec (or expressed as 33,000 ft-lb/min). This rate has become the definition of the standard unit called horsepower (see Equation 2.19).

$$\text{Horsepower (hp)} = \frac{\text{Power (ft-lb/min)}}{\text{33,000 ft-lb/min/hp}} \tag{2.19}$$

In water/wastewater treatment, the major use of horsepower calculation is in pumping stations. When used for this purpose, the horsepower calculation can be modified as shown below.

Water Horsepower

The amount of power required to move a given volume of water a specified total head is known as water horsepower (whp).

$$\text{Water horsepower (whp)} = \frac{\text{Pump rate (gpm)} \times \text{Total head (ft)} \times 8.34\ \text{lb/gal}}{\text{33,000 ft-lb/min/hp}} \tag{2.20}$$

■ **Example 2.103**

Problem: A pump must deliver 1210 gpm to a total head of 130 ft. What is the required water horsepower?

Solution:

$$\text{Required horsepower} = \frac{1210\ \text{gpm} \times 130\ \text{ft} \times 8.34\ \text{lb/gal}}{\text{33,000 ft-lb/min/hp}} = 40\ \text{whp}$$

Brake Horsepower

Brake horsepower (bhp) refers to the horsepower supplied to the pump from the motor. As power moves through the pump, additional horsepower is lost from slippage and friction of the shaft and other factors; thus, pump efficiencies range from about 50 to 85% and pump efficiency must be taken into account.

$$\text{Brake horsepower (bhp)} = \frac{\text{Water horsepower (whp)}}{\text{\% Efficiency of pump}} \tag{2.21}$$

■ **EXAMPLE 2.104**

Problem: Under the specified conditions, the pump efficiency is 73%. If the required water horsepower is 40 hp, what is the required brake horsepower?

Solution:

$$\text{Required horsepower} = \frac{40 \text{ whp}}{0.73} = 55 \text{ bhp}$$

Motor Horsepower

Motor horsepower (mhp) is the horsepower the motor must generate to produce the desired brake and water horsepower.

$$\text{Motor horsepower (mhp)} = \frac{\text{Brake horsepower (bhp)}}{\text{\% Efficiency of motor}} \tag{2.22}$$

■ **EXAMPLE 2.105**

Problem: A motor is 93% efficient. What is the required motor horsepower when the required brake horsepower is 49.0 bhp?

Solution:

$$\text{Motor horsepower} = \frac{49 \text{ bhp}}{0.93} = 53 \text{ mhp}$$

Electrical Power

Water/wastewater operators (especially senior operators) must occasionally make electrical power calculations, especially with regard to electrical energy required or consumed over a period of time. To accomplish this, horsepower is converted to electrical energy (kilowatts) and then multiplied by the hours of operation to obtain kilowatt-hours:

$$\text{Kilowatt-hours} = \text{Horsepower (hp)} \times 0.746 \text{ kW/hp} \times \text{Operating time (hr)} \tag{2.23}$$

■ **EXAMPLE 2.106**

Problem: A 60-hp motor operates at full load 12 hr/day, 7 days a week. How many kilowatts of energy does it consume per day?

Solution:

$$\text{Kilowatt-hours/day} = 60\text{ hp} \times 0.746\text{ kW/hp} \times 12\text{ hr/day} = 537\text{ kWh/day}$$

Given the cost per kilowatt-hour, the operator (or anyone else) can calculate the cost of power for any given period of operation.

$$\text{Cost} = \text{Power required/day} \times \text{kWh/day} \times \text{Days/period} \times \text{Cost/kWh} \quad (2.24)$$

■ **EXAMPLE 2.107**

Problem: A 60-hp motor requires 458 kWh/day. The pump is in operation every day. The current cost of electricity is $0.0328/kWh. What is the yearly electrical cost for this pump?

Solution:

$$\text{Cost} = 458\text{ kWh/day} \times 365\text{ days/year} \times \$0.0328\text{/kWh} = \$5{,}483.18$$

POWER FACTOR

Power factor is important because customers whose loads have low power factor require greater generation capacity than what is actually metered. This imposes a cost on the electric utility that is not otherwise recorded by the energy and demand charges. Two types of power make up the total or *apparent power* supplied by the electric utility. The first is *active* (also called *true*) *power*. Measured in kilowatts, it is the power used by the equipment to produce work. The second is *reactive power*. This is the power used to create the magnetic field necessary for induction devices to operate. It is measured in units of kilovolt-ampere reactive (kVAR).

- *Active power (true power) (P)*—Power that performs work, measured in watts (W)
- *Reactive power*—Power that does not perform work (sometimes called *wattless power*), measured in volt-ampere reactive (VAR)
- *Complex power*—The vector sum of the true and reactive power, measured in volt-ampere (VA)
- *Apparent power*—The magnitude of the complex power, measured in volt-ampere (VA)

The vector sum of the active and reactive power is complex power. Power factor is the ratio of the active power to the apparent power. The power factor of fully loaded induction motors ranges from 80 to 90%, depending on the type of motor and the speed of the motor. Power factor deteriorates as the load on the motor decreases. Other electrical devices such as space heaters and old fluorescent or high-discharge lamps also have poor power factor. Treatment plants have several motors, numerous lamps, and often electric heaters, which, combined, lower the facility's overall power factor.

Power factor may be leading or lagging. Voltage and current waveforms are in phase in a resistive alternating current (AC) circuit; however, reactive loads, such as induction motors, store energy in their magnetic fields. When this energy gets released back to the circuit it pushes the current and voltage waveforms out of phase. The current waveform then lags behind the voltage waveform.

Improving power factor is beneficial, as it improves voltage, decreases system losses, frees capacity to the system, and decreases power costs where fees for poor power factor are billed. Power factor can be improved by reducing the reactive power component of the circuit. Adding capacitors to an induction motor is perhaps the most cost-effective means to correct power factor as they provide reactive power. Synchronous motors are an alternative to capacitors for power factor correction.

Like induction motors, synchronous motors have stator windings that produce a rotating magnetic field; however, unlike the induction motor, the synchronous motor requires a separate source of DC from the field. It also requires special starting components. These include a salient-pole field with starting grid winding. The rotor of the conventional type of synchronous motor is essentially the same as that of the salient-pole AC generator. The stator windings of induction and synchronous motors are essentially the same.

In operation, the synchronous motor rotor locks into step with the rotating magnetic field and rotates at the same speed. If the rotor is pulled out of step with the rotating stator field, no torque is developed and the motor stops. Because a synchronous motor develops torque only when running at synchronous speed, it is not self-starting and hence requires some device to bring the rotor to synchronous speed. For example, the rotation of a synchronous motor may be started with a DC motor on a common shaft. After the motor is brought to synchronous speed, AC current is applied to the stator windings. The DC starting motor now acts as a DC generator, supplying DC field excitation for the rotor. The load then can be coupled to the motor.

Synchronous motors can be run at lagging, unity, or leading power factor by controlling their field excitation. When the field excitation voltage is decreased, the motor runs in lagging power factor. This condition is referred to as *under-excitation*. When the field excitation voltage is made equal to the rated voltage, the motor runs at unity power factor. The motor runs at leading power factor when the field excitation voltage is increased above the rated voltage. This condition is *over-excitation*. When over-excited, synchronous motors can provide system power factor correction.

The feasibility of adding capacitors depends on whether the electric utility charges for low power factor. Corrective measures are infrequently installed, as many electric utilities do not charge small customers for poor power factor but rather price it into the electrical rates as a cost of business. A cost evaluation is necessary to determine the type of correction equipment to use. The evaluation should include motor type, motor starter, exciter (for synchronous motors), capacitors, switching devices (if needed), efficiency, and power factor fees (IEEE, 1990). Manufacturers should be consulted before installing capacitors to reduced voltage solid-state starters and variable frequency drives, as there can be problems if they are not properly located and applied.

ADVANCED UNITS OF MEASUREMENT AND CONVERSIONS

When readers reach this point in the presentation and move into the material that follows, they may scratch their heads in wonder and ask: "Didn't we just cover this material in the earlier part of this chapter?"

The answer? Well, sort of. We did cover many of the items presented below but the reader should recall the basic thematic statement posted at the beginning of Chapter 1:

> The key to learning math can be summed up in one word: Repetition! Repetition! Repetition!

Thus, the following sections are partly based on providing the underlying structure, base, and foundation on which mathematical measurement and conversions are made (i.e., repetition), while the rest of the chapter covers additional areas that are becoming increasingly important to water or wastewater operators worldwide. Air pollution parameters and unit conversions are one example. Incinerating biosolids and eventual ash disposal can contribute to air pollution problems, along with soil contamination via ash disposal. In addition, greenhouse gas conversions units are covered in this chapter because biosolids are being applied as liquids and compost to soils as a viable amendment. This is especially the case in the Pacific Northwest where glaciated soils have resulted in nutrient-poor soils. The addition of treated biosolids to nutrient-poor forest areas has shown significant improvement in the growth rates of various timber types. Moreover, organic matter in biosolids improves soil structure, which reduces runoff and erosion. Soil organisms also benefit from nutrients in biosolids and the healthy soils it creates. Biosolids provide healthy, lush growth in the understory of forests which, in turn, provides more food and hiding cover for animals. We also address carbon footprint parameters and units and conversions associated with vehicle use at plants and other activities such as barbecuing during plant picnics and other plant social gatherings.

CONVERSION FACTORS AND SI UNITS

The units most commonly used by environmental engineering professionals are based on the complicated English system of weights and measures; however, bench work is usually based on the metric system, or International System of Units (SI), due to the convenient relationship between milliliters (mL), cubic centimeters (cm^3), and grams (g). The SI is a modernized version of the metric system established by international agreement. The metric system of measurement was developed during the French Revolution and was first promoted in the United States in 1866. In 1902, proposed congressional legislation requiring the U.S. government to use the metric system exclusively was defeated by a single vote. Although we use both systems in this text, the SI provides a logical and interconnected framework for all measurements in engineering, science, industry, and commerce. The metric system is much simpler to use than the existing English system, because all of its units of measurement are divisible by 10.

Before listing the various conversion factors commonly used in environmental engineering it is important to describe the prefixes commonly used in the SI system. These prefixes are based on the power 10. For example, a "kilo" means 1000 grams,

TABLE 2.5
SI Prefixes

Factor	Name	Symbol	Factor	Name	Symbol
10^{24}	Yotta	Y	10^{-1}	Deci	d
10^{21}	Zetta	Z	10^{-2}	Centi	c
10^{18}	Exa	E	10^{-3}	Milli	m
10^{15}	Peta	P	10^{-6}	Micro	µ
10^{12}	Tera	T	10^{-9}	Nano	n
10^{9}	Giga	G	10^{-12}	Pico	p
10^{6}	Mega	M	10^{-15}	Femto	f
10^{3}	Kilo	k	10^{-18}	Atto	a
10^{2}	Hecto	h	10^{-21}	Zepto	z
10^{1}	Deka	da	10^{-24}	Yocto	y

and a "centimeter" means 1/100 of 1 meter. The 20 SI prefixes used to form decimal multiples and submultiples of SI units are given in Table 2.5. Note that the kilogram is the only SI unit with a prefix as part of its name and symbol. Because multiple prefixes may not be used, in the case of the kilogram the prefix names of Table 2.5 are used with the unit name "gram" and the prefix symbols are used with the unit symbol "g." With this exception, any SI prefix may be used with any SI unit, including the degree Celsius and its symbol, °C.

■ **Example 2.108**

10^{-6} kg = 1 mg (1 milligram) is acceptable, but not 10^{-6} kg = 1 µkg (1 microkilogram).

DID YOU KNOW?

The Fibonacci sequence is the following sequence of numbers:

1, 1, 2, 3, 5, 8, 13, 21, 34, 55, 89, 144, …

Or, alternatively,

0.1, 1, 2, 3, 5, 8, 13, 21, 34, 55, 89, 144, …

Each term from the third one onward is *the sum of the previous two*. Another point to notice is that, if you divide each number in the sequence by the next number, beginning with the first, an interesting thing appears to be happening:

1/1 = 1, 1/2 = 0.5, 2/3 = 0.66666, 3/5 = 0.6,
5/8 = 0.625, 8/13 = 0.61538, 13/21 = 0.61904, …

Note that the first of these ratios appear to be converging to a number just a bit larger than 0.6.

■ Example 2.109

Consider the height of the Washington Monument. We may write it as 169,000 mm, 16,900 cm, 169 m, or 0.169 km, using the units of millimeter (prefix "milli," symbol "m"), centimeter (prefix "centi," symbol "c"), or kilometer (prefix "kilo," symbol "k").

Conversion Factor Tables

Conversion factors are given alphabetically in Table 2.6 and are listed by unit category in Table 2.7.

■ Example 2.110

Problem: Find degrees in Celsius of water at 72°F.

Solution:

$$°C = 5/9 \times (°F - 32) = 5/9 \times (72 - 32) = 22.2$$

Water/Wastewater Conversion Examples

■ Example 2.111

Convert cubic feet to gallons.

$$\text{Gallons (gal)} = \text{Cubic feet (ft}^3) \times 7.48 \text{ gal/ft}^3$$

Problem: How many gallons of biosolids can be pumped to a digester that has 3800 ft³ of volume available?

Solution:

$$\text{Gallons} = 3800 \text{ ft}^3 \times 7.48 \text{ gal/ft}^3 = 28{,}424 \text{ gal}$$

■ Example 2.112

Convert gallons to cubic feet.

$$\text{Cubic feet (ft}^3) = \text{Gallons} \div 7.48 \text{ gal/ft}^3$$

DID YOU KNOW?

Units and dimensions are not the same concepts. *Dimensions* are concepts such as time, mass, length, or weight. *Units* are specific cases of dimensions, such as hour, gram, meter, or pound. You can *multiply* and *divide* quantities with different units: 4 ft × 8 lb = 32 ft-lb, but you can *add* and *subtract* terms only if they have the same units. So, 5 lb + 8 kg = *no way!*

TABLE 2.6
Alphabetical List of Conversion Factors

Factor	Metric (SI) or English Conversions
°C	(5/9)[(°F) – 32°]
°F	(9/5)[(°C) + 32°]
1°C (expressed as an interval)	33.8°F = (9/5)(°F)
	1.8°R (degrees Rankine)
	1.0 K (degrees Kelvin)
1°F (expressed as an interval)	0.556°C = (5/9)°C
	1.0°R (degrees Rankine)
	0.556 K (degrees Kelvin)
1 atm (atmosphere)	1.013 bars
	10.133 N/cm^2 (newtons/square centimeter)
	33.90 ft of H_2O (feet of water)
	101.325 kPa (kilopascals)
	1,013.25 mbar (millibars)
	psia (pounds/square inch absolute)
	760 torr
	760 mmHg (millimeters of mercury)
1 bar	0.987 atm (atmosphere)
	1×10^6 dynes/cm^2 (dynes/square centimeter)
	33.45 ft of H_2O (feet of water)
	1×10^5 Pa (pascals)
	750.06 torr
	750.06 mmHg (millimeters of mercury)
1 Bq (becquerel)	1 radioactive disintegration/second
	2.7×10^{-11} Ci (curie)
	2.7×10^{-8} mCi (millicurie)
1 Btu (British thermal unit)	252 cal (calories)
	1055.06 J (joules)
	10.41 L-atm (liter-atmospheres)
	0.293 Wh (watt-hour)
1 cal (calorie)	3.97×10^{-3} Btu (British thermal unit)
	4.18 J (joules)
	0.0413 L-atm (liter-atmospheres)
	1.163×10^{-3} Wh (watt-hour)
1 Ci (curie)	3.7×10^{10} radioactive disintegrations/second
	3.7×10^{10} Bq (becquerels)
	1000 mCi (millicuries)
1 cm (centimeter)	0.0328 ft (foot)
	0.394 in. (inch)
	10,000 µm (microns, micrometers)
	100,000,000 Å = 10^8 Å (Ångstroms)
1 cm^2 (square centimeter)	1.076×10^{-3} ft^2 (square feet)
	0.155 in.^{2} (square inch)
	1×10^{-4} m^2 (square meter)

TABLE 2.6 (continued)
Alphabetical List of Conversion Factors

Factor	Metric (SI) or English Conversions
1 cm^3 (cc, cubic centimeter)	3.53×10^{-5} ft^3 (cubic feet)
	0.061 $in.^3$ (cubic inch)
	2.64×10^{-4} gal (gallon)
	52.18 L (liters)
	52.18 mL (milliliters)
1 day	24 hr (hours)
	1440 min (minutes)
	86,400 sec (seconds)
	0.143 wk (week)
	2.738×10^{-3} yr (year)
1 dyne	1×10^{-5} N (newton)
1 erg	1 dyn·cm (dyne-centimeter)
	1×10^{-7} J (joule)
	2.78×10^{-11} Wh (watt-hour)
1 eV (electron volt)	1.602×10^{-12} erg
	1.602×10^{-19} J (joule)
1 fps (foot per second)	1.097 kmph (kilometers/hour)
	0.305 mps (meter/second)
	0.01136 mph (mile/hour)
1 ft (foot)	30.48 cm (centimeters)
	12 in. (inches)
	0.3048 m (meter)
	1.65×10^{-4} NM (nautical mile)
	1.89×10^{-4} mi (statute mile)
1 ft^2 (square foot)	2.296×10^{-5} ac (acre)
	9.296 cm^2 (square centimeters)
	144 $in.^2$ (square inches)
	0.0929 m^2 (square meter)
1 ft^3 (cubic foot)	28.317 cm^3 (cc, cubic centimeters)
	1728 $in.^3$ (cubic inches)
	0.0283 m^3 (cubic meter)
	7.48 gal (gallons)
	28.32 L (liters)
	29.92 qt (quarts)
1 g (gram)	0.001 kg (kilogram)
	1000 mg (milligrams)
	1,000,000 ng = 10^6 ng (nanograms)
	2.205×10^{-3} lb (pound)
1 g/cm^3 (gram per cubic centimeter)	62.43 lb/ft^3 (pounds/cubic foot)
	0.0361 lb/$in.^3$ (pound/cubic inch)
	8.345 lb/gal (pounds/gallon)

TABLE 2.6 (continued)
Alphabetical List of Conversion Factors

Factor	Metric (SI) or English Conversions
1 gal (gallon)	3785 cm^3 (cc, cubic centimeters)
	0.134 ft^3 (cubic feet)
	231 $in.^3$ (cubic inches)
	3.785 L (liters)
1 Gy (gray)	1 J/kg (joule/kilogram)
	100 rad
	1 Sv (sievert), unless modified through division by an appropriate factor, such as *Q* or *N*
1 hp (horsepower)	745.7 J/sec (joules/second)
1 hr (hour)	0.0417 day
	60 min (minutes)
	3600 sec (seconds)
	5.95×10^{-3} wk (week)
	1.14×10^{-4} yr (year)
1 in. (inch)	2.54 cm (centimeters)
	1000 mil
1 $in.^3$ (cubic inch)	16.39 cm^3 (cc, cubic centimeters)
	16.39 mL (milliliters)
	5.79×10^{-4} ft^3 (cubic feet)
	1.64×10^{-5} m^3 (cubic meter)
	4.33×10^{-3} gal (gallon)
	0.0164 L (liter)
	0.55 fl oz. (fluid ounce)
1 inch of water	1.86 mmHg (millimeters of mercury)
	249.09 Pa (pascals)
	0.0361 psi (lb/$in.^2$)
1 J (joule)	9.48×10^{-4} Btu (British thermal unit)
	0.239 cal (calories)
	10,000,000 ergs = 1×10^7 ergs
	9.87×10^{-3} L atm (liter-atmospheres)
	1.0 N-m (newton-meter)
1 kcal (kilocalorie)	3.97 Btu (British thermal units)
	1000 cal (calories)
	4186.8 J (joules)
1 kg (kilogram)	1000 g (grams)
	2205 lb (pounds)
1 km (kilometer)	3280 ft (feet)
	0.54 NM (nautical mile)
	0.6214 mi (statute mile)

TABLE 2.6 (continued)
Alphabetical List of Conversion Factors

Factor	Metric (SI) or English Conversions
1kW (kilowatt)	56.87 Btu/min (British thermal units per minute)
	1.341 hp (horsepower)
	1000 J/sec (joules per second)
1 L (liter)	1000 cm^3 (cc, cubic centimeters)
	1 dm^3 (cubic decimeter)
	0.0353 ft^3 (cubic feet)
	61.02 $in.^3$ (cubic inches)
	0.264 gal (gallon)
	1000 mL (milliliters)
	1.057 qt (quarts)
1 lb (pound)	453.59 g (grams)
	16 oz. (ounces)
1 lb/ft^3 (pound per cubic foot)	16.02 g/L (grams/liter)
1 lb/$in.^3$ (pound per cubic inch)	27.68 g/cm^3 (grams/cubic centimeter)
	1728 lb/ft^3 (pounds/cubic feet)
1 m (meter)	1×10^{10} Å (Ångstroms)
	100 cm (centimeters)
	3.28 ft (feet)
	39.37 in. (inches)
	1×10^{-3} km (kilometer)
	1000 mm (millimeters)
	1,000,000 μm = 1×10^6 μm (micrometers)
	1×10^9 nm (nanometers)
1 m^2 (square meter)	10.76 ft^2 (square feet)
	1550 $in.^2$ (square inches)
1 m^3 (cubic meter)	1,000,000 cm^3 = 10^6 cm^3 (cc, cubic centimeters)
	33.32 ft^3 (cubic feet)
	61,023 $in.^3$ (cubic inches)
	264.17 gal (gallons)
	1000 L (liters)
1 mCi (millicurie)	0.001 Ci (curie)
	3.7×10^{10} radioactive disintegrations/second
	3.7×10^{10} Bq (becquerel)
1 mi (statute mile)	5280 ft (feet)
	1.609 km (kilometers)
	1609.3 m (meters)
	0.869 NM (nautical miles)
	1760 yd (yards)

TABLE 2.6 (continued)
Alphabetical List of Conversion Factors

Factor	Metric (SI) or English Conversions
1 mi^2 (square mile)	640 acres
	2.79×10^7 ft^2 (square feet)
	2.59×10^6 m^2 (square meters)
1 min (minute)	6.94×10^{-4} day
	0.0167 hr (hour)
	60 sec (seconds)
	9.92×10^{-5} wk (week)
	1.90×10^{-6} yr (year)
1 mmHg (mm of mercury)	1.316×10^{-3} atm (atmosphere)
	0.535 inch H_2O (inch of water)
	1.33 mb (millibars)
	133.32 Pa (pascals)
	1 torr
	0.0193 psia (pounds/square inch absolute)
1 mph (mile per hour)	88 fpm (feet/minute)
	1.61 kmph (kilometers/hour)
	0.447 mps (meters/second)
1 mps (meter per second)	196.9 fpm (feet/minute)
	3.6 kmph (kilometers/hour)
	2.237 mph (miles/hour)
1 N-m (newton-meter)	1.00 J (joule)
1 NM (nautical mile)	6076.1 ft (feet)
	1.852 km (kilometers)
	1.15 mi (statute miles)
	2025.4 yd (yards)
1 Pa (pascal)	9.87×10^{-6} atm (atmosphere)
	4.015×10^{-3} inch H_2O (inches of water)
	0.01 mb (millibar)
	7.5×10^{-3} mmHg (millimeters of mercury)
1 ppm (parts per million)	1.00 mL/m^3 (milliliter/cubic meter)
	1.00 mg/kg (milligram/kilogram)
1 psi (pounds/square inch)	0.068 atm (atmosphere)
	27.67 inch H_2O (inches of water)
	68.85 mb (millibars)
	51.71 mmHg (millimeters of mercury)
	6894.76 Pa (pascals)
1 qt (quart)	946.4 cm^3 (cc, cubic centimeters)
	57.75 in.^{3} (cubic inches)
	0.946 L (liter)

TABLE 2.6 (continued)
Alphabetical List of Conversion Factors

Factor	Metric (SI) or English Conversions
1 rad	100 ergs/g (ergs/gram)
	0.01 Gy (gray)
	1 rem, unless modified through division by an appropriate factor, such as *Q* or *N*
1 rem	1 rad, unless modified through division by an appropriate factor, such as *Q* or *N*
1 Sv (sievert)	1 Gy, unless modified through division by an appropriate factor, such as *Q* or *N*
1 torr	1.33 mb (millibar)
1 W (watt)	3.41 Btu/hr (British thermal units/hour)
	1.341×10^{-3} hp (horsepower)
	52.18 J/sec (joules/second)
1 wk (week)	7 days
	168 hr (hours)
	10,080 min (minutes)
	6.048×10^{5} sec (seconds)
	0.0192 yr (year)
1 Wh (watt-hour)	3.412 Btu (British thermal units)
	859.8 cal (calories)
	3600 J (joules)
	35.53 L-atm (liter-atmospheres)
1 yd^{3} (cubic yard)	201.97 gal (gallons)
	764.55 L (liters)
1 yr (year)	365.25 day
	8766 hr (hour)
	5.26×10^{5} min (minute)
	3.16×10^{7} sec (second)
	52.18 wk (week)

Problem: How many cubic feet of biosolids are removed when 14,000 gal are withdrawn?

Solution:

$$\text{Cubic feet} = 14{,}000 \text{ gal} \div 7.48 \text{ gal/ft}^3 = 1871.66 \text{ ft}^3$$

■ Example 2.113

Convert gallons to pounds.

$$\text{Pounds (lb)} = \text{Gallons} \times 8.34 \text{ lb/gal}$$

TABLE 2.7
Conversion Factors by Unit Category

Factor	Metric (SI) or English Conversions
	Units of Length
1 cm (centimeter)	0.0328 ft (foot)
	0.394 in. (inch)
	10,000 μm (microns, micrometers)
	100,000,000 Å = 10^8 Å (Ångstroms)
1 ft (foot)	30.48 cm (centimeters)
	12 in. (inches)
	0.3048 m (meter)
	1.65×10^{-4} NM (nautical mile)
	1.89×10^{-4} mi (statute mile)
1 in. (inch)	2.54 cm (centimeter)
	1000 mils
1 km (kilometer)	3280.8 ft (feet)
	0.54 NM (nautical mile)
	0.6214 mi (statute mile)
1 m (meter)	1×10^{10} Å (Ångstroms)
	100 cm (centimeters)
	3.28 ft (feet)
	39.37 in. (inches)
	1×10^{-3} km (kilometer)
	1000 mm (millimeters)
	1,000,000 μm = 1×10^6 μm (microns, micrometers)
	1×10^9 nm (nanometers)
1 NM (nautical mile)	6076.1 ft (feet)
	1.852 km (kilometers)
	1.15 mi (statute miles)
	2025.4 yd (yards)
1 mi (statute mile)	5280 ft (feet)
	1.609 km (kilometers)
	1690.3 m (meters)
	0.869 NM (nautical mile)
	1760 yd (yards)
	Units of Area
1 cm^2 (square centimeter)	1.076×10^{-3} ft^2 (square feet)
	0.155 $in.^2$ (square inch)
	1×10^{-4} m^2 (square meter)
1 ft^2 (square foot)	2.296×10^{-5} ac (acre)
	929.03 cm^2 (square centimeters)
	144 $in.^2$ (square inches)
	0.0929 m^2 (square meter)
1 m^2 (square meter)	10.76 ft^2 (square feet)
	1550 $in.^2$ (square inches)

TABLE 2.7 (continued)
Conversion Factors by Unit Category

Factor	Metric (SI) or English Conversions
1 mi² (square mile)	640 acres
	2.79×10^7 ft² (square feet)
	2.59×10^6 m² (square meter)
	Units of Volume
1 cm³ (cubic centimeter)	3.53×10^{-5} ft³ (cubic feet)
	0.061 in.³ (cubic inch)
	2.64×10^{-4} gal (gallon)
	0.001 L (liter)
	1.00 mL (milliliter)
1 ft³ (cubic foot)	28,317 cm³ (cc, cubic centimeters)
	1728 in.³ (cubic inches)
	0.0283 m³ (cubic meter)
	7.48 gal (gallons)
	28.32 L (liters)
	29.92 qt (quarts)
1 in.³ (cubic inch)	16.39 cm³ (cc, cubic centimeters)
	16.39 mL (milliliters)
	5.79×10^{-4} ft³ (cubic feet)
	1.64×10^{-5} m³ (cubic meter)
	4.33×10^{-3} gal (gallon)
	0.0164 L (liter)
1 m³ (cubic meter)	1,000,000 cm³ = 10^6 cm³ (cc, cubic centimeters)
	35.31 ft³ (cubic feet)
	61,023 in.³ (cubic inches)
	264.17 gal (gallons)
	1000 L (liters)
1 yd³ (cubic yard)	201.97 gal (gallons)
	764.55 L (liters)
1 gal (gallon)	3785 cm³ (cc, cubic centimeters)
	0.134 ft³ (cubic feet)
	231 in.³ (cubic inches)
	3.785 L (liters)
1 L (liter)	1000 cm³ (cc, cubic centimeters)
	1 dm³ (cubic decimeter)
	0.0353 ft³ (cubic feet)
	61.02 in.³ (cubic inches)
	0.264 gal (gallon)
	1000 mL (milliliters)
	1.057 qt (quarts)
1 qt (quart)	946.4 cm³ (cc, cubic centimeters)
	57.75 in.³ (cubic inches)
	0.946 L (liter)

TABLE 2.7 (continued)
Conversion Factors by Unit Category

Factor	Metric (SI) or English Conversions
	Units of Mass
1 g (gram)	0.001 kg (kilogram)
	1000 mg (milligrams)
	1,000,000 mg = 10^6 ng (nanograms)
	2.205×10^{-3} lb (pounds)
	15.43 gr (grains)
	8.99×10^{13} J (joules)
1 kg (kilogram)	1000 g (grams)
	2.205 lb (pounds)
1 lb (pound)	453.59 g (grams)
	0.45359 kg (kilogram)
	16 oz. (ounces)
	Units of Time
1 day	24 hr (hours)
	1440 min (minutes)
	86,400 sec (seconds)
	0.143 wk (week)
	2.738×10^{-3} yr (year)
1 hr (hour)	0.0417 day
	60 min (minutes)
	3600 sec (seconds)
	5.95×10^{-3} wk (week)
	1.14×10^{-4} yr (year)
1 min (minute)	6.94×10^{-4} day
	0.0167 hr (hour)
	60 sec (seconds)
	9.92×10^{-5} wk (week)
	1.90×10^{-6} yr (year)
1 wk (week)	7 days
	168 hr (hours)
	10,080 min (minutes)
	6.048×10^{5} sec (seconds)
	0.0192 yr (year)
1 yr (year)	365.25 days
	8766 hr (hours)
	5.26×10^{5} min (minutes)
	3.16×10^{7} sec (seconds)
	52.18 wk (weeks)

TABLE 2.7 (continued)
Conversion Factors by Unit Category

Factor	Metric (SI) or English Conversions
	Units of the Measure of Temperature
°C	(5/9)[(°F) – 32°]
1°C (expressed as an interval)	33.8°F = (9/5)(°F)
	1.8°R (degrees Rankine)
	1.0 K (degree Kelvin)
°F Fahrenheit)	(9/5)[(°C) + 32°]
1°F (expressed as an interval)	0.556°C = (5/9)°C
	1.0°R (degree Rankine)
	0.556 K (degree Kelvin)
	Units of Force
1 dyne	1×10^{-5} N (newton)
1 N (newton)	1×10^{5} dyne
	Units of Work or Energy
1 Btu (British thermal unit)	252 cal (calories)
	1055.06 J (joules)
	10.41 L-atm (liter-atmospheres)
	0.293 Wh (watt-hour)
1 cal (calorie)	3.97×10^{-3} Btu (British thermal unit)
	4.18 J (joules)
	0.0413 L-atm (liter-atmosphere)
	1.163×10^{-3} Wh (watt-hour)
1 eV (electron volt)	1.602×10^{-12} erg
	1.602×10^{-19} J (joule)
1 erg	1 dyne-centimeter
	1×10^{-7} J (joule)
	2.78×10^{-11} Wh (watt-hour)
1 J (joule)	9.48×10^{-4} Btus (British thermal unit)
	0.239 cal (calorie)
	10,000,000 ergs = 1×10^{7} ergs
	9.87×10^{-3} L-atm (liter-atmosphere)
	1.00 N-m (newton-meter)
1 kcal (kilocalorie)	3.97 Btu (British thermal unit)
	1000 cal (calories)
	4186.8 J (joules)
1 kWh (kilowatt-hour)	3412.14 Btu (British thermal units)
	3.6×10^{6} J (joules)
	859.8 kcal (kilocalories)

TABLE 2.7 (continued)
Conversion Factors by Unit Category

Factor	Metric (SI) or English Conversions
1 N-m (newton-meter)	1.00 J (joule)
	2.78×10^{-4} Wh (watt-hour)
1 Wh (watt-hour)	3.412 Btu (British thermal units)
	859.8 cal (calories)
	3600 J (joules)
	35.53 L-atm (liter-atmospheres)
	Units of Power
1 hp (horsepower)	745.7 J/sec (joules/second)
1 kW (kilowatt)	56.87 Btu/min (British thermal units/minute)
	1.341 hp (horsepower)
	1000 J/sec (joules/second)
1 W (watt)	3.41 Btu/hr (British thermal units/hour)
	1.341×10^{-3} hp (horsepower)
	1.00 J/sec (joules/second)
	Units of Pressure
1 atm (atmosphere)	1.013 bars
	10.133 N/cm^2 (newtons/square centimeter)
	33.90 ft H_2O (feet of water)
	101.325 kPa (kilopascals)
	14.70 psia (pounds per square inch absolute)
	760 torr
	760 mmHg (millimeters of mercury)
1 bar	0.987 atm (atmosphere)
	1×10^6 dynes/cm^2 (dynes/square centimeter)
	33.45 ft H_2O (feet of water)
	1×10^5 Pa (pascal)
	750.06 torr
	750.06 mmHg (millimeters of mercury)
1 inch of water	1.86 mmHg (millimeters of mercury)
	249.09 Pa (pascals)
	0.0361 psi (lb/in.^{2})
1 mmHg (millimeter of mercury)	1.316×10^{-3} atm (atmosphere)
	0.535 in H_2O (inches of water)
	1.33 mb (millibars)
	133.32 Pa (pascals)
	1 torr
	0.0193 psia (pounds per square inch absolute)
1 pascal	9.87×10^{-6} atm (atmosphere)
	4.015×10^{-3} inch H_2O (inches of water)
	0.01 mb (millibar)
	7.5×10^{-3} mmHg (millimeters of mercury)

TABLE 2.7 (continued)
Conversion Factors by Unit Category

Factor	Metric (SI) or English Conversions
1 psi (pounds per square inch)	0.068 atm (atmosphere)
	27.67 inch H_2O (inches of water)
	68.85 mb (millibars)
	51.71 mmHg (millimeters of mercury)
	6894.76 Pa (pascals)
1 torr	1.33 mb (millibars)
Units of Velocity or Speed	
1 fps (foot per second)	1.097 kmph (kilometers/hour)
	0.305 mps (meters/second)
	0.01136 mph (miles/hour)
1 mps (meters per second)	196.9 fpm (feet/minute)
	3.6 kmph (kilometers/hour)
	2.237 mph (miles/hour)
1 mph (miles per hour)	88 fpm (feet/minute)
	1.61 kmph (kilometers/hour)
	0.447 mps (meters/second)
Units of Density	
1 g/cm^3 (grams per cubic centimeter)	62.43 lb/ft^3 (pounds/cubic foot)
	0.0361 lb/in.^{3} (pound/cubic inch)
	8.345 lb/gal (pounds/gallon)
1 lb/ft^3 (pounds/cubic foot)	16.02 g/L (grams/liter)
1 lb/in.^{2} (pounds/cubic inch)	27.68 g/cm^3 (grams/cubic centimeter)
	1.728 lb/ft^3 (pounds/cubic foot)
Units of Concentration	
1 ppm (parts/million-volume)	1.00 mL/m^3 (milliliter/cubic meter)
1 ppm (wt)	1.00 mg/kg (milligram/kilogram)
Radiation and Dose Related Units	
1 Bq (becquerel)	1 radioactive disintegration/second
	2.7×10^{-11} Ci (curie)
	2.7×10^{-8} (millicurie)
1 Ci (curie)	3.7×10^{10} radioactive disintegrations/second
	3.7×10^{10} Bq (becquerels)
	1000 mCi (millicuries)
1 Gy (gray)	1 J/kg (joule/kilogram)
	100 rad
	1 Sv, unless modified through division by an appropriate factor, such as *Q* or *N*

TABLE 2.7 (continued)
Conversion Factors by Unit Category

Factor	Metric (SI) or English Conversions
1 mCi (millicurie)	0.001 Ci (curie)
	3.7×10^{10} radioactive disintegrations/second
	3.7×10^{10} Bq (becquerels)
1 rad	100 ergs/g (ergs/gram)
	0.01 Gy (gray)
	1 rem, unless modified through division by an appropriate factor, such as *Q* or *N*
1 rem	1 rad, unless modified through division by an appropriate factor, such as *Q* or *N*
1 Sv (sievert)	1 Gy, unless modified through division by an appropriate factor, such as *Q* or *N*

Problem: If 1850 gal of solids are removed from the primary settling tank, how many pounds of solids are removed?

Solution:

$$\text{Pounds} = 1850 \text{ gal} \times 8.34 \text{ lb/gal} = 15{,}429 \text{ lb}$$

■ Example 2.114

Convert pounds to gallons.

$$\text{Gallons (gal)} = \text{Pounds (lb)} \div 8.34 \text{ lb/gal}$$

Problem: How many gallons of water are required to fill a tank that holds 8500 lb of water?

Solution:

$$\text{Gallons} = 8500 \text{ lb} \div 8.34 \text{ lb/gal} = 1019.18 \text{ gal}$$

■ Example 2.115

Convert milligrams per liter to pounds.

> ***Note:*** For plant operations, concentrations in milligrams per liter or parts per million determined by laboratory testing must be converted to quantities of pounds, kilograms, pounds per day, or kilograms per day.

$$\text{Pounds (lb)} = \text{Concentration (mg/L)} \times \text{Volume (MG)} \times 8.34 \text{ lb/MG/mg/L}$$

Problem: The solids concentration in the aeration tank is 2500 mg/L. The aeration tank volume is 0.90 MG. How many pounds of solids are in the tank?

Solution:

$$\text{Pounds} = 2500 \text{ mg/L} \times 0.90 \text{ MG} \times 8.34 \text{ lb/MG/mg/L} = 18{,}765 \text{ lb}$$

■ Example 2.116

Convert milligrams per liter to pounds per day.

$$\text{Pounds/day (lb/day)} = \text{Concentration (mg/L)} \times \text{Flow (MGD)} \times 8.34 \text{ lb/MG/mg/L}$$

Problem: How many pounds of solids are discharged per day when the plant effluent flow rate is 4.85 MGD and the effluent solids concentration is 25 mg/L?

Solution:

$$\text{Pounds/day} = 25 \text{ mg/L} \times 4.85 \text{ MGD} \times 8.34 \text{ lb/MG/mg/L} = 1011.23 \text{ lb/day}$$

■ Example 2.117

Convert milligrams per liter to kilograms per day.

$$\text{Kilograms/day} = \text{Concentration (mg/L)} \times \text{Volume (MG)} \times 3.785 \text{ L/gal}$$

Problem: The effluent contains 25 mg/L of BOD_5. How many kilograms per day of BOD_5 are discharged when the effluent flow rate is 9.3 MGD?

Solution:

$$\text{Kilograms/day} = 25 \text{ mg/L} \times 9.3 \text{ MGD} \times 3.785 \text{ L/gal} = 880.01 \text{ kg/day}$$

■ Example 2.118

Convert pounds to milligrams per liter.

$$\text{Concentration (mg/L)} = \frac{\text{Quantity (lb)}}{\text{Volume (MG)} \times 8.34 \text{ lb/MG/mg/L}}$$

Problem: An aeration tank contains 89,900 lb of solids. The volume of the aeration tank is 4.40 MG. What is the concentration of solids in the aeration tank in milligrams per liter?

Solution:

$$\text{Concentration} = \frac{89{,}000 \text{ lb}}{4.40 \text{ MG} \times 8.34 \text{ lb/MG/mg/L}} = 2449.6 \text{ mg/L}$$

■ Example 2.119

Convert pounds per day to milligrams per liter.

$$\text{Concentration (mg/L)} = \frac{\text{Quantity (lb)}}{\text{Volume (MGD)} \times 8.34 \text{ lb/MG/mg/L}}$$

Problem: The disinfection process uses 4800 lb/day of chlorine to disinfect a flow of 25 MGD. What is the concentration of chlorine applied to the effluent?

Solution:

$$\text{Concentration} = \frac{4800\text{ lb}}{25\text{ MGD} \times 8.34\text{ lb/MG/mg/L}} = 23\text{ mg/L}$$

■ Example 2.120

Convert pounds to flow in million gallons per day.

$$\text{Flow (MGD)} = \frac{\text{Quantity (lb)}}{\text{Concentration (mg/L)} \times 8.34\text{ lb/MG/mg/L}}$$

Problem: 9600 lb of solids must be removed from the activated biosolids process per day. The waste activated biosolids concentration is 7690 mg/L. How many million gallons per day of waste activated biosolids must be removed?

Solution:

$$\text{Flow} = \frac{9600\text{ lb}}{7690\text{ mg/L} \times 8.34\text{ lb/MG/mg/L}} = 0.15\text{ MGD}$$

■ Example 2.121

Convert million gallons per day to gallons per minute.

$$\text{Flow (gpm)} = \frac{\text{Flow (MGD)} \times 1{,}000{,}000\text{ gal/MG}}{1440\text{ min/day}}$$

Problem: The current flow rate is 5.50 MGD. What is the flow rate in gallons per minute?

Solution:

$$\text{Flow} = \frac{5.50\text{ MGD)} \times 1{,}000{,}000\text{ gal/MG}}{1440\text{ min/day}} = 3819.4\text{ gpm}$$

■ Example 2.122

Convert million gallons per day to gallons per day.

$$\text{Flow (gpd)} = \text{Flow (MGD)} \times 1{,}000{,}000\text{ gal/MG}$$

Problem: The influent meter reads 28 MGD. What is the current flow rate in gallons per day?

Solution:

$$\text{Flow} = 28 \text{ MGD} \times 1{,}000{,}000 \text{ gal/MG} = 28{,}800{,}000 \text{ gpd}$$

■ **EXAMPLE 2.123**

Convert million gallons per day to cubic feet per second.

$$\text{Flow (cfs)} = \text{Flow (MGD)} \times 1.55 \text{ cfs/MGD}$$

Problem: The flow rate entering the grit channel is 2.90 MGD. What is the flow rate in cubic feet per second?

Solution:

$$\text{Flow} = 2.90 \text{ MGD} \times 1.55 \text{ cfs/MGD} = 4.50 \text{ cfs}$$

■ **EXAMPLE 2.124**

Convert gallons per minute to million gallons per day.

$$\text{Flow (MGD)} = \frac{\text{Flow (gpm)} \times 1440 \text{ min/day}}{1{,}000{,}000 \text{ gal/MG}}$$

Problem: The flow meter indicates that the current flow rate is 1475 gpm. What is the flow rate in million gallons per day?

Solution:

$$\text{Flow} = \frac{1475 \text{ gpm} \times 1440 \text{ min/day}}{1{,}000{,}000 \text{ gal/MG}} = 2.12 \text{ MGD}$$

■ **EXAMPLE 2.125**

Convert gallons per day to million gallons per day.

$$\text{Flow (MGD)} = \frac{\text{Flow (gpd)}}{1{,}000{,}000 \text{ gal/MG}}$$

Problem: A totalizing flow meter indicates that 33,400,900 gal of wastewater have entered the plant in the past 24 hr. What is the flow rate in million gallons per day?

Solution:

$$\text{Flow} = \frac{33{,}400{,}900 \text{ gpd}}{1{,}000{,}000 \text{ gal/MG}} = 33.4 \text{ MGD}$$

■ EXAMPLE 2.126

Convert flow in cubic feet per second to million gallons per day.

$$\text{Flow (MGD)} = \text{Flow (cfs)} \div 1.55 \text{ cfs/MG}$$

Problem: The flow in a channel is determined to be 3.90 cfs. What is the flow rate in million gallons per day?

Solution:

$$\text{Flow} = 3.90 \text{ cfs} \div 1.55 \text{ cfs/MG} = 2.5 \text{ MGD}$$

■ EXAMPLE 2.127

Problem: The water in a tank weighs 670 lb. How many gallons does it hold?

Solution: Water weighs 8.34 lb/gal; therefore,

$$670 \text{ lb} \div 8.34 \text{ lb/gal} = 80.3 \text{ gal}$$

■ EXAMPLE 2.128

Problem: A liquid chemical weighs 64 lb/ft³. How much does a 5-gal can of it weigh?

Solution: Solve for specific gravity, determine lb/gal, and multiply by 5.

$$\text{Specific gravity} = \frac{\text{Weight of chemical}}{\text{Weight of water}} = \frac{64 \text{ lb/ft}^3}{62.4 \text{ lb/ft}^3} = 1.03$$

$$1.03 = \frac{\text{Weight of chemical}}{8.34 \text{ lb/gal}}$$

$$1.03 \times 8.34 \text{ lb/gal} = \text{Weight of chemical} = 8.59 \text{ lb/gal}$$

$$8.59 \text{ lb/gal} \times 5 \text{ gal} = 43 \text{ lb}$$

■ EXAMPLE 2.129

Problem: A wooden piling with a diameter of 16 in. and a length of 20 ft weighs 50 lb/ft³. If it is inserted vertically into a body of water, what vertical force is required to hold it below the water surface?

Solution: If this piling had the same weight as water, it would rest just barely submerged. Find the difference between its weight and that of the same volume of water—that is the weight needed to keep it down:

$$62.4 \text{ lb/ft}^3 \text{ (water)} - 50.0 \text{ lb/ft}^3 \text{ (piling)} = 12.4 \text{ lb/ft}^3$$

$$\text{Volume of piling} = 0.785 \times (1.33)^2 \times 20 \text{ ft} = 27.8 \text{ ft}^3$$

$$12.4 \text{ lb/ft}^3 \times 27.8 \text{ ft}^3 = 344.7 \text{ lb}$$

■ **Example 2.130**

Problem: A liquid chemical with a specific gravity of 1.22 is being pumped at a rate of 50 gpm. How many pounds per day are being delivered by the pump?

Solution: Solve for pounds pumped per minute, then change to pounds/day.

$$8.34 \text{ lb/gal water} \times 1.22 = 10.2 \text{ lb/gal liquid}$$

$$50 \text{ gal/min} \times 10.2 \text{ lb/gal} = 510 \text{ lb/min}$$

$$510 \text{ lb/min} \times 1440 \text{ min/day} = 734{,}400 \text{ lb/day}$$

■ **Example 2.131**

Problem: A cinder block weighs 80 lb in air. When it is immersed in water, it weighs 40 lb. What are the volume and specific gravity of the cinder block?

Solution: The cinder block displaces 30 lb of water; solve for cubic feet of water displaced (equivalent to volume of cinder block).

$$30 \text{ lb water displaced} \div 62.4 \text{ lb/ft}^3 = 0.48 \text{ ft}^3 \text{ water displaced}$$

Cinder block volume is 0.48 ft^3, which weighs 80 lb; thus,

$$80 \text{ lb} \div 0.48 \text{ ft}^3 = 166.7 \text{ lb/ft}^3$$

$$\text{Specific gravity} = \frac{\text{Density of cinder block}}{\text{Density of water}} = \frac{166.7 \text{ lb/ft}^3}{62.4 \text{ lb/ft}^3} = 2.67$$

Temperature Conversion Examples

The two common methods for making temperature conversions are

- $°C = 5/9(°F - 32)$
- $°F = 9/5(°C) + 32$

■ **Example 2.132**

Problem: At a temperature of 4°C, water is at its greatest density. What is that temperature in degrees Fahrenheit?

Solution:

$$(9/5 \times °C) + 32 = (9/5 \times 4) + 32 = 7.2 + 32 = 39.2°F$$

The difficulty arises when one tries to recall these formulas from memory. Probably the easiest way to recall these important formulas is to remember these basic steps for both Fahrenheit and Celsius conversions:

1. Add 40°.
2. Multiply by the appropriate fraction (5/9 or 9/5).
3. Subtract 40°.

Obviously, the only variable in this method is the choice of 5/9 or 9/5 in the multiplication step. To make the proper choice, you must be familiar with the two scales. The freezing point of water is 32° on the Fahrenheit scale and 0° on the Celsius scale. The boiling point of water is 212° on the Fahrenheit scale and 100° on the Celsius scale.

> ***Note:*** At the same temperature, higher numbers are associated with the Fahrenheit scale and lower numbers with the Celsius scale. This important relationship helps you decide whether to multiply by 5/9 or 9/5.

Now look at a few conversion problems to see how the three-step process works.

■ Example 2.133

Problem: Convert 200°F to Celsius.

Solution: Using the three-step process, we proceed as follows:

1. Add 40°:

$$200° + 40° = 240°$$

2. 240° must be multiplied by either 5/9 or 9/5. Because the conversion is to the Celsius scale, we will be moving to a number *smaller* than 240. Through reason and observation, obviously, if 240 were multiplied by 9/5, the result would be almost the same as multiplying by 2, which would double 240 rather than make it smaller. If we multiply by 5/9, the result will be about the same as multiplying by 1/2, which would cut 240 in half. Because in this problem we wish to move to a smaller number, we should multiply by 5/9:

$$5/9 \times 240° = 133.3°\text{C}$$

3. Now subtract 40°:

$$133.3°\text{C} - 40.0°\text{C} = 93.3°\text{C}$$

Therefore, 200°F = 93.3°C.

■ Example 2.134

Problem: Convert 24°C to Fahrenheit.

Solution: Using the three-step process, we proceed as follows:

1. Add 40°:

$$24° + 40° = 64°$$

2. Because we are converting from Celsius to Fahrenheit, we are moving from a smaller to a larger number, and 9/5 should be used in the multiplication:

$$9/5 \times 64° = 115.2°$$

3. Subtract 40°:

$$115.2° - 40° = 75.2°$$

Thus, 22°C = 75.2°F.

Obviously, knowing how to make these temperature conversion calculations is useful. However, in practical *in situ* or non-*in situ* operations, you may wish to use a temperature conversion table as shown in Table 2.8.

TABLE 2.8
Temperature Conversion Table

°C	°F	°C	°F	°C	°F	°C	°F	°C	°F	°C	°F	°C	°F
–40	–40.0	–10	14.0	20	68.0	50	122.0	80	176.0	110	230.0	140	284.0
–39	–38.2	–9	15.8	21	69.8	51	123.8	81	177.8	111	231.8	141	285.8
–38	–36.4	–8	17.6	22	71.6	52	125.6	82	179.6	112	233.6	142	287.6
–37	–34.6	–7	19.4	23	73.4	53	127.4	83	181.4	113	235.4	143	289.4
–36	–32.8	–6	21.2	24	75.2	54	129.2	84	183.2	114	237.2	144	291.2
–35	–31.0	–5	23.0	25	77.0	55	131.0	85	185.0	115	239.0	145	293.0
–34	–29.2	–4	24.8	26	78.8	56	132.8	86	186.8	116	240.8	146	294.8
–33	–27.4	–3	26.6	27	80.6	57	134.6	87	188.6	117	242.6	147	296.6
–32	–25.6	–2	28.4	28	82.4	58	136.4	88	190.4	118	244.4	148	298.4
–31	–23.8	–1	30.2	29	84.2	59	138.2	89	192.2	119	246.2	149	300.2
–30	–22.0	0	32.0	30	86.0	60	140.0	90	194.0	120	248.0	150	302.0
–29	–20.2	1	33.8	31	87.8	61	141.8	91	195.8	121	249.8	151	303.8
–28	–18.4	2	35.6	32	89.6	62	143.6	92	197.6	122	251.6	152	305.6
–27	–16.6	3	37.4	33	91.4	63	145.4	93	199.4	123	253.4	153	307.4
–26	–14.8	4	39.2	34	93.2	64	147.2	94	201.2	124	255.2	154	309.2
–25	–13.0	5	41.0	35	95.0	65	149.0	95	203.0	125	257.0	155	311.0
–24	–11.2	6	42.8	36	96.8	66	150.8	96	204.8	126	258.8	156	312.8
–23	–9.4	7	44.6	37	98.6	67	152.6	97	206.6	127	260.6	157	314.6
–22	–7.6	8	46.4	38	100.4	68	154.4	98	208.4	128	262.4	158	316.4
–21	–5.8	9	48.2	39	102.2	69	156.2	99	210.2	129	264.2	159	318.2
–20	–4.0	10	50.0	40	104.0	70	158.0	100	212.0	130	266.0	160	320.0
–19	–2.2	11	51.8	41	105.8	71	159.8	101	213.8	131	267.8	161	321.8
–18	–0.4	12	53.6	42	107.6	72	161.6	102	215.6	132	269.6	162	323.6
–17	1.4	13	55.4	43	109.4	73	163.4	103	217.4	133	271.4	163	325.4
–16	3.2	14	57.2	44	111.2	74	165.2	104	219.2	134	273.2	164	327.2
–15	5.0	15	59.0	45	113.0	75	167.0	105	221.0	135	275.0	165	329.0
–14	6.8	16	60.8	46	114.8	76	168.8	106	222.8	136	276.8	166	330.8
–13	8.6	17	62.6	47	116.6	77	170.6	107	224.6	137	278.6	167	332.6
–12	10.4	18	64.4	48	118.4	78	172.4	108	226.4	138	280.4	168	334.4
–11	12.2	19	66.2	49	120.2	79	174.2	109	228.2	139	282.2	169	336.2

Conversion Factors for Air Pollution Measurements

The USEPA Part 503 regulation establishes requirements for sewage biosolids-only incinerators. The rule covers the biosolids feed, the furnace itself, the operation of the furnace and the exhaust gases from the stack. It is the exhaust gases, potential air pollutants, that are of interest to us here. Simply put, wastewater plant managers and operators and responsible persons in charge of the proper operation biosolids incinerators must be aware of not only regulatory compliance requirements but also air pollution control parameters and associated air pollution measurements.

Air pollutant emissions are commonly stated in metric system whole numbers. If possible, the reported units should be the same as those that are actually being measured. For example, weight should be recorded in grams, and volume of air should be recorded in cubic meters. When the analytical system is calibrated in one unit, the emissions should be reported in the same units of the calibration standard. For example, if a gas chromatograph is calibrated with a 1-ppm standard of toluene in air, then the emissions monitored by the system should also be reported in ppm. Finally, if the emission standard is defined in a specific unit, the monitoring system should be selected to monitor in that unit. Tables 2.9 and 2.10 illustrate the conversions of various volumes to attain 1 part per million (ppm) and conversions for parts per million in proportion and percent.

The preferred reporting units for the following types of emissions are

Nonmethane organic and volatile organic compound emissions	ppm, ppb
Semi-volatile organic compound emissions	$\mu g/m^3$, mg/m^3
Particulate matter (TSP/PM-10) emissions	$\mu g/m^3$
Metal compound emissions	ng/m^3

Conversion from ppm to $\mu g/m^3$

Often, the environmental practitioner must be able to convert from ppm to $\mu g/m^3$. Following is an example of how one would perform that conversion using sulfur dioxide (SO_2) as the monitored constituent.

TABLE 2.9
Conversion for Various Volumes to Attain One Part Per Million

Amount of Active Ingredient	Unit of Volume	Parts per Million
2.71 pounds	Acre-foot	1 ppm
1.235 grams	Acre-foot	1 ppm
1.24 kilograms	Acre-foot	1 ppm
0.0283 grams	Cubic foot	1 ppm
1 milligram	Liter	1 ppm
8.34 pounds	Million gallons	1 ppm
1 gram	Cubic meter	1 ppm
0.0038 grams	Gallon	1 ppm
3.8 grams	Thousand gallons	1 ppm

TABLE 2.10
Conversion for Parts per Million in Proportion and Percent

Parts per Million	Proportion	Percent	Parts per Million	Proportion	Percent
0.1	1:10,000,000	0.00001	25.0	1:40,000	0.0025
0.5	1:2,000,000	0.00005	50.0	1:20,000	0.005
1.0	1:1,000,000	0.0001	100.0	1:10,000	0.01
2.0	1:500,000	0.0002	200.0	1:5,000	0.02
3.0	1:333,333	0.0003	250.0	1:4,000	0.025
5.0	1:200,000	0.0005	500.0	1:2,000	0.05
7.0	1:142,857	0.0007	1550.0	1:645	0.155
10.0	1:100,000	0.001	5000.0	1:200	0.5
15.0	1:66,667	0.0015	10,000.0	1:100	1.0

The expression "parts per million" is without dimensions; that is, no units of weight or volume are specifically designated. Using the format of other units, the expression may be written as

$$\frac{\text{Parts}}{\text{Million parts}}$$

"Parts" are not defined. If cubic centimeters replace parts, we obtain:

$$\frac{\text{Cubic centimeter}}{\text{Million cubic centimeters}}$$

Similarly, we might write pounds per million pounds, tons per million tons, or liters per million liters. In each expression, identical units of weight of volume appear in both the numerator and denominator and may be canceled out, leaving a dimensionless term. An analog of parts per million is the more familiar term "percent." Percent can be written as

$$\frac{\text{Parts}}{\text{Hundred parts}}$$

To convert from parts per million by volume (μL/L) to μg/m³ at standard temperature (25°C) and standard pressure (760 mmHg), known as STP, it is necessary to know the molar volume at the given temperature and pressure and the molecular weight of the pollutant. At 25°C and 760 mmHg, 1 mole of any gas occupies 24.46 L.

■ Example 2.135

Problem: 2.5 ppm by volume of sulfur dioxide (SO_2) was reported as the atmospheric concentration. What is this concentration in micrograms (μg) per cubic meter (m³) at 25°C and 760 mmHg? What is the concentration in μg/m³ at 37°C and 752 mmHg?

Note: This example problem points out the need for reporting temperature and pressure when the results are present on a weight-to-volume basis.

Solution: Let parts per million equal μL/L, then 2.5 ppm = 2.5 μL/L. The molar volume at 25°C and 760 mmHg is 24.46 L, and the molecular weight of SO_2 is 64.1 g/mole.

1. 25°C and 760 mmHg:

$$\frac{2.5\ \mu L}{L} \times \frac{1\ \mu mole}{24.46\ \mu L} \times \frac{64.1\ \mu g}{\mu mole} \times \frac{1000\ L}{m^3} = \frac{6.6 \times 10^3\ \mu g}{m^3} \text{ at STP}$$

2. 37°C and 752 mmHg:

$$24.46\ \mu L \times \frac{310°K}{298°K} \times \frac{760\ mmHg}{752\ mmHg} = 25.72\ \mu L$$

$$\frac{2.5\ \mu L}{L} \times \frac{1\ \mu mole}{25.72\ \mu L} \times \frac{64.1\ \mu g}{\mu mole} \times \frac{1000\ L}{m^3} = \frac{6.2 \times 10^3\ \mu g}{m^3} \text{ at STP}$$

Conversion Tables for Common Air Pollution Measurements

To assist the environmental engineer in converting from one set of units to another, the following conversion factors for common air pollution measurements and other useful information are provided. The conversion tables provide factors for

- Atmospheric gases
- Atmospheric pressure
- Gas velocity
- Concentration
- Atmospheric particulate matter

Following is a list of conversions from ppm to μg/m³ (at 25°C and 760 mmHg) for several common air pollutants. Also see Tables 2.11 to 2.15.

- ppm SO_2 × 2620 = μg/m³ SO_2 (sulfur dioxide)
- ppm CO × 1150 = μg/m³ CO (carbon monoxide)
- ppm NH_3 × 696 = μg/m³ NH_3 (ammonia)
- ppm CO_2 × 1800 = μg/m³ CO_2 (carbon dioxide)
- ppm NO × 1230 = μg/m³ NO (nitrogen oxide)
- ppm NO_2 × 1880 = μg/m³ NO_2 (nitrogen dioxide)
- ppm O_3 × 1960 = μg/m³ O_3 (ozone)
- ppm CH_4 × 655 = μg/m³ CH_4 (methane)
- ppm CH_4 × 655 = μg/m³ CH_4 (methane)
- ppm CH_3SH × 2000 = μg/m³ CH_3SH (methyl mercaptan)
- ppm C_3H_8 × 1800 = μg/m³ C_3H_8 (propane)

TABLE 2.11
Atmospheric Gases

To Convert from	to	Multiply by
Milligram per cubic meter (mg/m^3)	Micrograms per cubic meter ($\mu m/m^3$)	1000.0
	Micrograms per liter ($\mu g/L$)	1.0
	ppm by volume (20°C)	24.04/molecular weight of gas
	ppm by weight	0.8347
	Pounds per cubic foot (lb/ft^3)	62.43×10^{-9}
Micrograms per cubic foot ($\mu m/ft^3$)	Milligrams per cubic foot (mg/ft^3)	0.001
	Micrograms per liter ($\mu g/L$)	0.001
	ppm by volume (20° C)	0.02404/molecular weight of gas
	ppm by weight	834.7×10^{-6}
	Pounds per cubic foot (lb/ft^3)	62.43×10^{-12}
Micrograms/liter ($\mu m/L$)	Milligrams per cubic meter (mg/m^3)	1.0
	Micrograms per cubic meter ($\mu g/m^3$)	1000.0
	ppm by volume (20°C)	24.04/molecular weight of gas
	ppm by weight	0.8347
	Pounds per cubic ft (lb/ft^3)	62.43×10^{-9}
ppm by volume (20°C)	Milligrams per cubic meter (mg/m^3)	Molecular weight of gas/24.04
	Micrograms per cubic meter ($\mu g/m^3$)	Molecular weight of gas/0.02404
	Micrograms per liter ($\mu g/L$)	Molecular weight of gas/24.04
	ppm by weight	Molecular weight of gas/28.8
	Pounds per cubic ft (lb/ft^3)	Molecular weight of gas/385.1×10^6
ppm by weight	Milligrams per cubic meter (mg/m^3)	1.198
	Micrograms per cubic meter ($\mu g/m^3$)	1.198×10^3
	Micrograms per liter ($\mu g/L$)	1.198
	ppm by volume (20°C)	28.8/molecular weight of gas
	Pounds per cubic ft (lb/ft^3)	7.48×10^{-6}
Pounds per cubic foot (lb/ft^3)	Milligrams per cubic meter (mg/m^3)	16.018×10^6
	Micrograms per cubic meter ($\mu g/m^3$)	16.018×10^9
	Micrograms per liter ($\mu g/L$)	16.018×10^6
	ppm by volume (20°)	385.1×10^6/molecular weight of gas
	ppm by weight	133.7×10^3

- ppm $C_3H_8 \times 1.8$ = mg/m^3 C_3H_8 (propane)
- ppm $F^- \times 790$ = $\mu g/m^3$ F^- (fluoride)
- ppm $H_2S \times 1400$ = $\mu g/m^3$ H_2S (hydrogen sulfide)
- ppm HCHO × 1230 = $\mu g/m^3$ HCHO (formaldehyde)

Soil Test Results Conversion Factors

Note: Because biosolids are often land-applied to agricultural fields or are land-filled, wastewater biosolids managers must be familiar with the USEPA's 503 rules for land application of biosolids.

TABLE 2.12
Atmospheric Pressure

To Convert from	to	Multiply by
Atmospheres	Millimeters of mercury	760.0
	Inches of mercury	29.92
	Millibars	1013.2
Millimeters of mercury	Atmospheres	1.316×10^{-3}
	Inches of mercury	39.37×10^{-3}
	Millibars	1.333
Inches of mercury	Atmospheres	0.03333
	Millimeters of mercury	25.4005
	Millibars	33.35
Millibars	Atmospheres	0.000987
	Millimeters of mercury	0.75
	Inches of mercury	0.30
	Sampling Pressures	
Millimeters of mercury	Inches of water (60°C)	0.5358
Inches of mercury	Inches of water (60°C)	13.609
Inches of water	Millimeters of mercury (0°C)	1.8663
	Inches of mercury (0°C)	73.48×10^{-2}

The USEPA's definition of land application includes all forms of applying bulk or bagged biosolids to land for beneficial uses at agronomic rates (rates designed provide the amount of nitrogen needed by the crop or vegetation grown on the land while minimizing the amount that passes below the root zone). These beneficial use practices include application to agricultural land such as fields used for the production

TABLE 2.13
Velocity

To Convert from	to	Multiply by
Meters/second (m/sec)	Kilometers/hour (km/hr)	3.6
	Feet/second (fps)	3.281
	Miles/hour (mph)	2.237
Kilometers/hour (km/hr)	Meters/second (m/sec)	0.2778
	Feet/second (fps)	0.9113
	Miles/hour (mph)	0.6241
Feet/hour (ft/hr)	Meters/second (m/sec)	0.3048
	Kilometers/hour (km/hr)	1.0973
	Miles/hour (mph)	0.6818
Miles/hour (mph)	Meters/second (m/sec)	0.4470
	Kilometers/hour (km/hr)	1.6093
	Feet/second (fps)	1.4667

TABLE 2.14
Atmospheric Particulate Matter

To Convert from	to	Multiply by
Milligrams/cubic meter (mg/m^3)	Grams/cubic foot (g/ft^3)	283.2×10^{-6}
	Grams/cubic meter (g/m^3)	0.001
	Micrograms/cubic meter ($\mu g/m^3$)	1000.0
	Micrograms/cubic foot ($\mu g/ft^3$)	28.32
	Pounds/1000 cubic feet ($lb/1000\ ft^3$)	62.43×10^{-6}
Grams/cubic foot (g/ft^3)	Milligrams/cubic meter (mg/m^3)	35.3145×10^3
	Grams/cubic meter (g/m^3)	35.314
	Micrograms/cubic meter ($\mu g/m^3$)	35.314×10^3
	Micrograms/cubic foot ($\mu g/ft^3$)	1.0×10^6
	Pounds/1000 cubic feet ($lb/1000\ ft^3$)	2.2046

of food, feed and fiber crops, pasture and range land; non-agricultural land such as forests; public contact sites such as parks and golf courses; disturbed lands such as mine spills, construction sites, and gravel pits; and home lawns and gardens. The sale or give away of biosolids products (such as composted or heat dried products) is addressed under land application, as is land application of domestic septage.

TABLE 2.15
Concentration

To Convert from	to	Multiply by
Grams/cubic meter (g/m^3)	Milligrams/cubic meter (mg/m^3)	1000.0
	Grams/cubic foot (g/ft^3)	0.02832
	Micrograms/cubic foot ($\mu g/ft^3$)	1.0×10^6
	Pounds/1000 cubic feet ($lb/1000\ ft^3$)	0.06243
Micrograms/cubic meter ($\mu g/m^3$)	Milligrams/cubic meter (mg/m^3)	0.001
	Grams/cubic foot (g/ft^3)	28.43×10^{-9}
	Grams/cubic meter (g/m^3)	1.0×10^{-6}
	Micrograms/cubic foot ($\mu g/ft^3$)	0.02832
	Pounds/1000 cubic feet ($lb/1000\ ft^3$)	62.43×10^{-9}
Micrograms/cubic foot ($\mu g/ft^3$)	Milligrams/cubic meter (mg/m^3)	35.314×10^{-3}
	Grams/cubic foot (g/ft^3)	1.0×10^{-6}
	Grams/cubic meter (g/m^3)	35.314×10^{-6}
	Micrograms/cubic foot ($\mu g/ft^3$)	35.314
	Pounds/1000 cubic feet ($lb/1000\ ft^3$)	2.2046×10^{-6}
Pounds/1000 cubic feet ($lb/1000\ ft^3$)	Milligrams/cubic meter (mg/m^3)	16.018×10^3
	Grams/cubic foot (g/ft^3)	0.35314
	Micrograms/cubic meter ($\mu g/m^3$)	16.018×10^6
	Grams/cubic meter (g/m^3)	16.018
	Micrograms/cubic foot ($\mu g/ft^3$)	353.14×10^2

TABLE 2.16
Soil Test Conversion Factors

Soil Sample Depth (inches)	Multiply ppm by
3	1
6	2
7	2.33
8	2.66
9	3
10	3.33
12	4

To ensure full compliance with the USEPA's biosolids land application rules, biosolids managers must conduct various analytical tests. Soil test results can be converted from parts per million (ppm) to pounds per acre by multiplying ppm by a conversion factor based on the depth to which the soil was sampled. Because a slice of soil 1 acre in area and 3 inches deep weighs approximately 1 million pounds, the conversion factors given in Table 2.16 can be used.

Greenhouse Gas Emission Numbers to Equivalent Units and Carbon Sequestration

This section describes the calculations used to convert greenhouse gas emission numbers—an area of increasing concern for environmental practitioners—into different types of equivalent units.

Electricity Reductions (Kilowatt-Hours)

The USEPA's Greenhouse Gas Equivalencies Calculator uses the Emissions & Generation Resource Integrated Database (eGRID) of U.S. annual non-baseload CO_2 output emission rates to convert reductions of kilowatt-hours into avoided units of carbon dioxide emissions. Most users of the Equivalencies Calculator who seek equivalencies for electricity-related emissions want to know equivalencies for emissions reductions due to energy efficiency or renewable energy programs. These programs are not generally assumed to affect baseload emissions (the emissions from power plants that run all the time), but rather non-baseload generation (power plants that are brought online as necessary to meet demand). For that reason, the Equivalencies Calculator uses a non-baseload emissions rate (USEPA, 2012a).

STANDARD CONVERSIONS FOR MANUAL CALCULATIONS

Volume	Weight	Length
1 gal = 3.78 L	1 lb = 453 g or 0.453 kg	1 in. = 2.54 cm
1 L = 0.26 gal	1 kg = 2.2 lb	1 cm = 0.39 in.
1 tsp = 5 mL		3.28 ft = 1 m

Emission Factor

$$7.0555 \times 10^{-4} \text{ metric tons } CO_2\text{/kWh}$$

Note: This calculation does not include any greenhouse gases other than CO_2, and it does not include line losses.

Passenger Vehicles per Year

Passenger vehicles are defined as two-axle, four-tire vehicles, including passenger cars, vans, pickup trucks, and sport/utility vehicles. In 2010, the weighted average combined fuel economy of cars and light trucks was 21.6 miles per gallon (FHWA, 2010). The average number of vehicle miles traveled in 2010 was 11,489 miles per year. In 2010, the ratio of carbon dioxide emissions to total greenhouse gas emissions (including carbon dioxide, methane, and nitrous oxide, all expressed as carbon dioxide equivalents) for passenger vehicles was 0.985 (USEPA, 2013b). The amount of carbon dioxide emitted per gallon of motor gasoline burned was 8.92×10^{-3} metric tons.

To determine annual greenhouse gas emissions per passenger vehicle, the following methodology was used: The amount of vehicle miles traveled (VMT) was divided by average gas mileage to determine gallons of gasoline consumed per vehicle per year. The number of gallons of gasoline consumed was multiplied by carbon dioxide per gallon of gasoline to determine carbon dioxide emitted per vehicle per year. Carbon dioxide emissions were then divided by the ratio of carbon dioxide emissions to total vehicle greenhouse gas emissions to account for vehicle methane and nitrous oxide emissions.

Calculation

Due to rounding, performing the calculations given in the equations below may not return the exact results shown.

(8.92×10^{-3} metric tons CO_2 per gal gasoline) × (11,489 VMT car/truck average) × (1/21.6 miles per gal car/truck average) × [(1 CO_2, CH_4, and N_2O)/0.985 CO_2] = 4.8 metric tons CO_2 emissions per vehicle per year.

Gallons of Gasoline Consumed

To obtain the number of grams of CO_2 emitted per gallon of gasoline combusted, the heat content of the fuel per gallon is multiplied by the kg CO_2 per heat content of the fuel. The average heat content per gallon of gasoline is 0.125 mmBtu/gal and the average emissions per heat content of gasoline is 71.35 kg CO_2/mmBtu (USEPA, 2012b). The fraction oxidized to CO_2 is 100% (IPCC, 2006).

Calculation

Due to rounding, performing the calculations given in the equations below may not return the exact results shown.

(0.125 mmBtu/gal) × (71.35 kg CO_2 per mmBtu) × (1 metric ton/1000 kg) = 8.92×10^{-3} metric tons CO_2 per gal of gasoline.

Therms of Natural Gas

Carbon dioxide emissions per therm are determined by multiplying heat content times the carbon coefficient times the fraction oxidized times the ratio of the molecular weight of carbon dioxide to that of carbon (44/12). The average heat content of natural gas is 0.1 mmBtu per therm, and the average carbon coefficient of natural gas is 14.47 kg carbon per mmBtu (USEPA, 2012b). The fraction oxidized to CO_2 is 100% (IPCC, 2006).

> ***Note:*** When using this equivalency, please keep in mind that it represents the CO_2 equivalency for natural gas burned as a fuel, not natural gas released to the atmosphere. Direct methane emissions released to the atmosphere (without burning) are about 21 times more powerful than CO_2 in terms of their warming effect on the atmosphere.

Calculation

Due to rounding, performing the calculations given in the equations below may not return the exact results shown.

(0.1 mmBtu/1 therm) × (14.47 kg C per mmBtu) × (44 g CO_2 per 12 g C) × (1 metric ton/1000 kg) = 0.005 metric tons CO_2 per therm.

Barrels of Oil Consumed

Carbon dioxide emissions per barrel of crude oil are determined by multiplying heat content times the carbon coefficient times the fraction oxidized times the ratio of the molecular weight of carbon dioxide to that of carbon (44/12). The average heat content of crude oil is 5.80 mmBtu per barrel, and the average carbon coefficient of crude oil is 20.31 kg carbon per mmBtu (USEPA, 2012b). The fraction oxidized to CO_2 is 100% (IPCC, 2006).

Calculation

Due to rounding, performing the calculations given in the equations below may not return the exact results shown.

(5.80 mmBtu/barrel) × (20.31 kg C per mmBtu) × (44 g CO_2 per 12 g C) × (1 metric ton/1000 kg) = 0.43 metric tons CO_2 per barrel.

Tanker Trucks Filled with Gasoline

Carbon dioxide emissions per barrel of gasoline are determined by multiplying the heat content times the carbon dioxide coefficient times the fraction oxidized times the ratio of the molecular weight of carbon dioxide to that of carbon (44/12). A barrel equals 42 gallons. A typical gasoline tanker truck contains 8500 gallons. The average heat content of conventional motor gasoline is 0.125 mmBtu/gal, and the average carbon coefficient of motor gasoline is 71.35 kg CO_2 (USEPA, 2012a). The fraction oxidized to CO_2 is 100% (IPCC, 2006).

Calculation

Due to rounding, performing the calculations given in the equations below may not return the exact results show.

(0.125 mmBtu/gal) × (71.35 kg CO_2 per mmBtu) × (1 metric ton/1000 kg) = 8.92 × 10^{-3} metric tons CO_2 per gallon.

(8.92 × 10^{-3} metric tons CO_2 per gallon) × (8500 gal per tanker truck) = 75.82 metric tons CO_2 per tanker truck.

Number of Tree Seedlings Grown for 10 Years

Forest application of biosolids is extensively practiced in the Pacific Northwest. Thus, we need to be concerned about the carbon footprint involved. A medium-growth coniferous tree, planted in an urban setting and allowed to grow for 10 years sequesters 23.2 lb of carbon. This estimate is based on the following assumptions:

- Medium-growth coniferous trees are raised in a nursery for one year until they become 1 inch in diameter at 4.5 feet above the ground (the size of tree purchased in a 15-gallon container).
- The nursery-grown trees are then planted in a suburban/urban setting; the trees are not densely planted.
- The calculation takes into account "survival factors" developed by the U.S. Department of Energy. For example, after 5 years (1 year in the nursery and 4 in the urban setting), the probability of survival is 68%; after 10 years, the probability declines to 59%. For each year, the sequestration rate (in pounds per tree) is multiplied by the survival factor to yield a probability-weighted sequestration rate. These values are summed over the 10-year period, beginning from the time of planting, to derive the estimate of 23.2 lb of carbon per tree.

Please note the following caveats to these assumptions:

- Although most trees take 1 year in a nursery to reach the seedling stage, trees grown under different conditions and trees of certain species may take longer—up to 6 years.
- Average survival rates in urban areas are based on broad assumptions, and the rates will vary significantly depending upon site conditions.
- Carbon sequestration depends on growth rate, which varies by location and other conditions.
- This method estimates only direct sequestration of carbon and does not include the energy savings that result from buildings being shaded by urban tree cover.

To convert to units of metric tons CO_2 per tree, multiply by the ratio of the molecular weight of carbon dioxide to that of carbon (44/12) and the ratio of metric tons per pound (1/2204.6).

Calculation

Due to rounding, performing the calculations given in the equations below may not return the exact results show.

(23.2 lb C per tree) × (44 units CO_2 ÷ 12 units C) × (1 metric ton ÷ 2204.6 lb) = 0.039 metric ton CO_2 per urban tree planted.

Acres of U.S. Forests Storing Carbon for One Year

Growing forests accumulate and store carbon. Through the process of photosynthesis, trees remove CO_2 from the atmosphere and store it as cellulose, lignin, and other compounds. The rate of accumulation is equal to growth minus removals (i.e., harvest for the production of paper and wood) minus decomposition. In most U.S. forests, growth exceeds removals and decomposition, so the amount of carbon stored nationally is increasing overall.

Calculation for U.S. Forests

The *Inventory of U.S. Greenhouse Gas Emissions and Sinks* (USEPA, 2012b) provides data on the net change in forest carbon stocks and forest area. Net changes in carbon attributed to harvested wood products are not included in the calculation.

Annual net change in carbon stocks per area in year n = (Carbon stocks$_{(t+1)}$ – Carbon stocks$_t$) ÷ (Area of land remaining in the same land-use category)

1. Determine the carbon stock change between years by subtracting carbon stocks in year t from carbon stocks in year $(t + 1)$. (This includes carbon stocks in the above-ground biomass, below-ground biomass, dead wood, litter, and soil organic carbon pools.)
2. Determine the annual net change in carbon stocks (i.e., sequestration) per area by dividing the carbon stock change in U.S. forests from step 1 by the total area of U.S. forests remaining in forests in year $(n + 1)$ (i.e., the area of land that did not change land-use categories between the time periods).

DID YOU KNOW?

Forest land in the United States includes land that is at least 10% stocked with trees of any size, or, in the case of stands dominated by certain western woodland species for which stocking parameters are not available, at least 5% crown cover by trees of any size. Timberland is defined as unreserved productive forest land producing or capable of producing crops of industrial wood. Productivity is at a minimum rate of 20 ft^3 of industrial wood per acre per year. The remaining portion of forest land is classified as "reserved forest land," which is forest withdrawn from timber use by statute or regulation, or "other forest land," which includes forests on which timber is growing at a rate less than 20 ft^3 per acre per year (Smith et al., 2010).

Applying these calculations to data developed by the USDA Forest Service for the *Inventory of U.S. Greenhouse Gas Emissions and Sinks* yields a result of 150 metric tons of carbon per hectare (or 61 metric tons of carbon per acre)* for the carbon stock density of U.S forests in 2010, with an annual net change in carbon stock per area in 2010 of 0.82 metric tons of carbon sequestered per hectare per year (or 0.33 metric tons of carbon sequestered per acre per year). These values include carbon in the five forest pools of above-ground biomass, below-ground biomass, deadwood, litter, and soil organic carbon, and they are based on state-level Forest Inventory and Analysis (FIA) data. Forest carbon stocks and carbon stock change are based on the stock difference methodology and algorithms described by Smith et al. (2010).

Conversion Factors for Carbon Sequestered Annually by One Acre of Average U.S. Forest

Due to rounding, performing the calculations given in the equations below may not return the exact results shown. In the following calculation, negative values indicate carbon sequestration.

(–0.33 metric ton C per acre/year) × (44 units CO_2 ÷ 12 units C) = –1.22 metric ton CO_2 sequestered annually by one acre of average U.S. forest.

Note that this is an estimate for "average" U.S. forests in 2010 (i.e., for U.S. forests as a whole in 2010). Significant geographical variations underlie the national estimates, and the values calculated here might not be representative of individual regions of states. To estimate carbon sequestered for additional acres in one year, simply multiply the number of acres by 1.22 metric tons CO_2 per acre/year. From 2000 to 2010, the average annual sequestration per area was 0.73 metric tons C per hectare/year (or 0.30 metric tons C per acre/year) in the United States, with a minimum value of 0.36 metric tons C per hectare/year (or 0.15 metric tons C per acre/year) in 2000, and a maximum value of 0.83 metric tons C per hectare/year (or 0.34 metric tons C per acre/year) in 2006.

Acres of U.S. Forest Preserved from Conversion to Cropland

The carbon stock density of U.S. forests in 2010 was 150 metric tons of carbon per hectare (or 61 metric tons of carbon per acre) (USEPA, 2012b). This estimate is composed of the five carbon pools of above-ground biomass (52 metric tons C per hectare), below-ground biomass (10 metric tons C per hectare), dead wood (9 metric tons C per hectare), litter (17 metric tons per C hectare), and soil organic carbons (62 metric tons C per hectare).

The *Inventory of U.S. Greenhouse Gas Emissions and Sinks* estimates soil carbon stock changes using U.S.-specific equations and data from the USDA Natural Resource Inventory and the CENTURY biogeochemical model (USEPA, 2012b). When calculating carbon stock changes in biomass due to conversion from forestland to cropland, the IPCC guidelines indicate that the average carbon stock change is equal to the carbon stock change due to removal of biomass from the

* 1 hectare = 10,000 m²; 100 m by 100 m; 2.47 acres.

outgoing land use (i.e., forestland) plus the carbon stocks from one year of growth in the incoming land use (i.e., cropland), or the carbon in biomass immediately after the conversion minus the carbon in biomass prior to the conversion plus the carbon stocks from one year of growth in the incoming land use (i.e., cropland) (IPCC, 2006). The carbon stock in annual cropland biomass after 1 year is 5 metric tons carbon per hectare, and the carbon content of dry above-ground biomass is 45% (IPCC, 2006). Therefore, the carbon stock in cropland after 1 year of growth is estimated to be 2.25 metric tons carbon per hectare (or 0.91 metric tons carbon per acre).

The averaged reference soil carbon stock (for high-activity clay, low-activity clay, and sandy soils for all climate regions in the United States) is 40.83 metric tons carbon per hectare (USEPA, 2012b). Carbon stock change in soils is time dependent, with a default time period for transition between equilibrium soil organ carbon values of 20 years for mineral soils in cropland systems (IPCC, 2006). Consequently, it is assumed that the change in equilibrium mineral soil organic carbon will be annualized over 20 years to represent the annual flux. The IPCC (2006) guidelines indicate that there are insufficient data to provide a default approach or parameters to estimate carbon stocks in perennial cropland.

Calculations for Converting U.S. Forests to U.S. Cropland

- ***Annual change in biomass carbon stocks on land converted to other land-use category:***

$$\Delta C_B = \Delta C_G + C_{Conversion} - \Delta C_L$$

where

ΔC_B = Annual change in carbon stocks in biomass due to growth on land converted to another land-use category (i.e., 2.25 metric tons C per hectare).

ΔC_G = Annual increase in carbon stocks in biomass due to growth on land converted to another land-use category (i.e., 2.25 metric tons C per hectare).

$C_{Conversion}$ = Initial change in carbon stocks in biomass on land converted to another land-use category; the sum of the carbon stocks in above-ground, below-ground, deadwood, and litter biomass (–88.47 metric tons C per hectare). Immediately after conversion from forestland to cropland, biomass is assumed to be zero, as the land is cleared of all vegetation before planting crops.

ΔC_L = Annual decrease in biomass stocks due to losses from harvesting, fuel wood gathering, and disturbances on land converted to other land-use category (assumed to be zero).

Therefore, $\Delta C_B = \Delta C_G + C_{Conversion} - \Delta C_L = -86.22$ metric tons carbon per hectare per year of biomass carbon stocks are lost when forestland is converted to cropland.

- ***Annual change in organic carbon stocks in mineral soils***

$$\Delta C_{Mineral} = (SOC_O - SOC_{(O-T)}) \div D$$

where

$\Delta C_{Mineral}$ = Annual change in carbon stocks in mineral soils.

SOC_O = Soil organic carbon stock in last year of inventory time period (i.e., 40.83 mt C per hectare).

$SOC_{(O-T)}$ = Solid organic carbon stock at beginning of inventory time period (i.e., 62 mt C per hectare).

D = Time dependence of stock change factors which is the default time period for transition between equilibrium SOC values (i.e., 20 years for cropland systems).

Therefore, $\Delta C_{Mineral}$ $(SPC_O - SOC_{(O-T)}) \div D = (40.83 - 62) \div 20 = -1.06$ metric tons C per hectare per year of soil organic C are lost. Consequently, the change in carbon density from converting forestland to cropland would be –86.22 metric tons of C per hectare per year of biomass plus –1.06 metric tons C per hectare per year of soil organic C, equaling a total loss of 87.28 metric tons C per hectare per year (or –35.32 metric tons C per acre per year). To convert to carbon dioxide, multiply by the ratio of the molecular weight of carbon dioxide to that of carbon (44/12), to yield a value of –320.01 metric tons CO_2 per hectare per year (or –129.51 metric tons CO_2 per acre per year).

Conversion Factor for Carbon Sequestered Annually by One Acre of Forest Preserved from Conversion to Cropland

Due to rounding, performing the calculations given in the equations below may not return the exact results shown. Negative values indicate CO_2 that is *not* emitted.

(–35.32 metric tons C per acre per year) × (44 units CO_2 ÷ 12 units C) = –129.51 metric tons CO_2 per acre per year.

To estimate CO_2 not emitted when an acre of forest is preserved form conversion to cropland, simply multiply the number of acres of forest not converted by –129.51 metric tons CO_2 per acre per year. Note that this calculation method assumes that all of the forest biomass is oxidized during clearing (i.e., one of the burned biomass remains as charcoal or ash). Also note that this estimate only includes mineral soil carbon stocks, as most forests in the contiguous United States are growing on mineral soils. In the case of mineral soil forests, soil carbon stocks could be replenished or even increased, depending on the starting stocks, how the agricultural lands are managed, and the time frame over which lands are managed.

Propane Cylinders Used for Home Barbecues

Propane is 81.7% carbon. The fraction oxidized is 100% (IPCC, 2006; USEPA, 2012b). Carbon dioxide emissions per pound of propane were determined by multiplying the weight of propane in a cylinder times the carbon content percentage times the fraction oxidized times the ratio of the molecular weight of carbon dioxide to that of carbon (44/12). Propane cylinders vary with respect to size; for the purpose of this equivalency calculation, a typical cylinder for home use was assumed to contain 18 pounds of propane.

Calculation

Due to rounding, performing the calculations given in the equations below may not return the exact results shown.

(18 lb propane per cylinder) × (0.817 lb C per lb propane) × (0.4536 kg/lb) × (44 kg CO_2 per 12 kg C) × (1 metric ton/1000 kg) = 0.024 metric tons CO_2 per cylinder.

Railcars of Coal Burned

The average heat content of coal in 2009 was 27.56 mmBtu per metric ton. The average carbon coefficient of coal in 2009 was 25.34 kg carbon per mmBtu (USEPA, 2011). The fraction oxidized to CO_2 is 100% (IPCC, 2006). Carbon dioxide emissions per ton of coal were determined by multiplying heat content times the carbon coefficient times the fraction oxidized times the ratio of the molecular weight of carbon dioxide to that of carbon (44/12). The amount of coal in an average railcar was assumed to be 100.19 short tons, or 90.89 metric tons (Hancock and Sreekanth, 2001).

Calculation

Due to rounding, performing the calculations given in the equations below may not return the exact results shown.

(27.56 mmBtu/metric ton coal) × (25.34 kg C per mmBtu) × (44g CO_2 per 12g C) × (90.89 metric tons coal per railcar) × (1 metric ton/1000 kg) = 232.74 metric tons CO_2 per railcar.

Tons of Waste Recycled Instead of Landfilled

To develop the conversion factor for recycling rather than landfilling waste, emission factors from the USEPA's Waste Reduction Model (WARM) were used (USEPA, 2012b). These emission factors were developed following a life-cycle assessment methodology using estimation techniques developed for national inventories of greenhouse gas emissions. According to WARM, the net emission reduction from recycling mixed recyclables (e.g., paper, metals, plastics), compared with a baseline in which the materials are landfilled, is 0.73 metric tons of carbon equivalent per short ton. This factor was then converted to metric tons of carbon dioxide equivalent by multiplying by 44/12, the molecular weight ratio of carbon dioxide to carbon.

Calculation

Due to rounding, performing the calculation given in the equation below may not return the exact results show.

(0.73 metric tons of carbon equivalent per ton) × (44 g CO_2 per 12 g C) = 2.67 metric tons CO_2 equivalent per ton of waste recycled instead of landfilled.

Coal-Fired Power Plant Emissions for One Year

In 2009, a total of 457 power plants used coal to generate at least 95% of their electricity (USEPA, 2012a). These plants emitted 1,614,625,638.1 metric tons of CO_2 in 2009. Carbon dioxide emissions per power plant were calculated by dividing the total emissions from power plants whose primary source of fuel was coal by the number of power plants.

Calculation

Due to rounding, performing the calculations given in the equations below may not return the exact results shown.

(1,614,625,638.1 metric tons of CO_2) × (1/457 power plants) = 3,533,098 metric tons CO_2 per power plant.

REFERENCES AND RECOMMENDED READING

FHWA. (2010). *Highway Statistics 2010*. Washington, DC: Office of Highway Policy Information, Federal Highway Administration, U.S. Department of Transportation (http://www.fhwa.dot.gov/policyinformation/statistics/2010/index.cfm).

Hancock, K. and Sreekanth, A. (2001). Conversion of weight of freight to number of railcars. *Transport. Res. Rec.*, 1768: 1–10.

IPCC. (2006). *2006 IPCC Guidelines for National Green House Gas Inventories*. Geneva, Switzerland: Intergovernmental Panel on Climate Change.

Rishel, J.B. (2001). Wire-to-water efficiency of pumping systems. *ASHRAE J.*, 43(4): 40–46.

Smith, J.L., Heath, L., and Nichols, M. (2010). *U.S. Forest Carbon Calculation Tool User's Guide: Forestland Carbon Stocks and Net Annual Stock Change*, General Technical Report NRS-13 revised. St. Paul, MN: U.S. Department of Agriculture Forest Service, Northern Research Station.

USEIA. (1998). *Method for Calculating Carbon Sequestration by Trees in Urban and Suburban Settings*. Washington, DC: U.S. Energy Information Administration.

USEIA. (2009). *2009 Residential Energy Consumption Survey*, Table CE2.6: Fuel Expenditures Totals and Averages, U.S. Homes. Washington, DC: U.S. Energy Information Administration.

USEPA. (2011). *Inventory of U.S. Greenhouse Gas Emissions and Sinks (MMT CO_2 Equivalents): Fast Facts 1990–2009*. Washington, DC: U.S. Environmental Protection Agency (http://www.epa.gov/climatechange/emissions/usinventoryreport.html).

USEPA. (2012a). *eGrid2012 Version 1.0 Year 2009 Summary Tables*. Washington, DC: U.S. Environmental Protection Agency (http://www.epa.gov/cleanenergy/documents/egridzips/eGRID2012V1_0_year09_SummaryTables.pdf).

USEPA. (2012b). *Inventory of U.S. Greenhouse Gas Emissions and Sinks: 1990–2011*. Washington, DC: U.S. Environmental Protection Agency (http://www.epa.gov/climatechange/ghgemissions/usinventoryreport.html).

USEPA. (2013). *Waste Reduction Model (WARM)*. Washington, DC: U.S. Environmental Protection Agency (http://epa.gov/epawaste/conserve/tools/warm/index.html).

3 Sequence of Operations

PEDMAS: Parentheses, Exponents, Multiplication, Division, Addition, Subtraction, *or* Please Excuse My Dear Aunt Sally

Mathematical operations such as addition, subtraction, multiplication, and division are usually performed in a certain order or sequence. Typically, multiplication and division operations are done prior to addition and subtraction operations. In addition, mathematical operations are also generally performed from left to right using this hierarchy. The use of parentheses is also common to set apart operations that should be performed in a particular sequence.

Note: It is assumed that the reader has a fundamental knowledge of basic arithmetic and math operations; thus, the purpose of the following section is to provide a brief review of the mathematical concepts and applications frequently employed by wastewater operators.

SEQUENCE OF OPERATIONS RULES

Rule 1

In a series of additions, the terms may be placed in any order and grouped in any way; thus, 4 + 3 = 7 and 3 + 4 = 7; (4 + 3) + (6 + 4) = 17, (6 + 3) + (4 + 4) = 17, and [6 + (3 + 4)] + 4 = 17.

Rule 2

In a series of subtractions, changing the order or the grouping of the terms may change the result; thus, 100 – 30 = 70, but 30 – 100 = –70, and (100 – 30) – 10 = 60, but 100 – (30 – 10) = 80.

Rule 3

When no grouping is given, the subtractions are performed in the order written, from left to right; thus, 100 – 30 – 15 – 4 = 51 (by steps, it would be 100 – 30 = 70, 70 – 15 = 55, 55 – 4 = 51).

Rule 4

In a series of multiplications, the factors may be placed in any order and in any grouping; thus, [(2 × 3) × 5] × 6 = 180 and 5 × [2 × (6 × 3)] = 180.

Rule 5

In a series of divisions, changing the order or the grouping may change the result; thus, $100 \div 10 = 10$ but $10 \div 100 = 0.1$, and $(100 \div 10) \div 2 = 5$ but $100 \div (10 \div 2) = 20$. Again, if no grouping is indicated, the divisions are performed in the order written, from left to right; thus, $100 \div 10 \div 2$ is understood to mean $(100 \div 10) \div 2$.

Rule 6

In a series of mixed mathematical operations, the convention is as follows: Whenever no grouping is given, multiplications and divisions are to be performed in the order written, then additions and subtractions in the order written.

SEQUENCE OF OPERATIONS EXAMPLES

In a series of additions, the terms may be placed in any order and grouped in any way:

$$4 + 6 = 10 \text{ and } 6 + 4 = 10$$

$$(4 + 5) + (3 + 7) = 19, (3 + 5) + (4 + 7) = 19, \text{ and } [7 + (5 + 4)] + 3 = 19$$

In a series of subtractions, changing the order or the grouping of the terms may change the result:

$$100 - 20 = 80, \text{ but } 20 - 100 = -80$$

$$(100 - 30) - 20 = 50, \text{ but } 100 - (30 - 20) = 90$$

When no grouping is given, the subtractions are performed in the order written, from left to right:

$$100 - 30 - 20 - 3 = 47$$

or by steps:

$$100 - 30 = 70, 70 - 20 = 50, 50 - 3 = 47$$

In a series of multiplications, the factors may be placed in any order and in any grouping:

$$[(3 \times 3) \times 5] \times 6 = 270 \text{ and } 5 \times [3 \times (6 \times 3)] = 270$$

In a series of divisions, changing the order or the grouping may change the result:

$$100 \div 10 = 10, \text{ but } 10 \div 100 = 0.1$$

$$(100 \div 10) \div 2 = 5, \text{ but } 100 \div (10 \div 2) = 20$$

If no grouping is indicated, the divisions are performed in the order written—from left to right:

$$100 \div 5 \div 2 \text{ is understood to mean } (100 \div 5) \div 2$$

In a series of mixed mathematical operations, the rule of thumb is that, whenever no grouping is given, multiplications and divisions are to be performed in the order written, then additions and subtractions in the order written.

■ **EXAMPLE 3.1**

Problem: Perform the following mathematical operations to solve for the correct answer:

$$(2+4)+(2\times 6)+\left(\frac{6+2}{2}\right)=?$$

Solution: Mathematical operations are typically performed going from left to right within an equation and within sets of parentheses. Perform all math operations within the sets of parentheses first:

$$2+4=6$$

$$2\times 6=12$$

$$\frac{6+2}{2}=\frac{8}{2}=4$$

Note that the addition of 6 and 2 was performed prior to dividing. Now perform all math operations outside of the parentheses from left to right:

$$6 + 12 + 4 = 22$$

■ **EXAMPLE 3.2**

Problem: Solve the following equation:

$$(4 - 2) + (3 \times 3) - (15 \div 3) - 8 = ?$$

Solution: Perform math operations inside each set of parentheses:

$$4 - 2 = 2$$

$$3 \times 3 = 9$$

$$15 \div 3 = 5$$

Perform addition and subtraction operations from left to right:

$$2 + 9 - 5 - 8 = -2$$

There may be cases where several operations will be performed within multiple sets of parentheses. In these cases we must perform all operations within the innermost set of parentheses first and move outward. We must continue to observe the hierarchical rules throughout the problem. Brackets, [ ], may indicate additional sets of parentheses.

■ **EXAMPLE 3.3**

Problem: Solve the following equation:

$$[2 \times (3 + 5) - 5 + 2] \times 3 = ?$$

Solution: Perform operations in the innermost set of parentheses

$$3 + 5 = 8$$

Rewrite the equation:

$$[2 \times 8 - 5 + 2] \times 3 = ?$$

Perform multiplication prior to addition and subtraction within the bracket.

$$[16 - 5 + 2] \times 3 = [11 + 2] \times 3 = [13] \times 3 = 13 \times 3 = 39$$

■ **EXAMPLE 3.4**

Problem: Solve the following equation:

$$7 + [2(3 + 1) - 1] \times 2 = ?$$

Solution:

$$7 + [2(4) - 1] \times 2 = 7 + [8 - 1] \times 2 = 7 + [7] \times 2 = 7 + 14 = 21$$

■ **EXAMPLE 3.5**

Problem: Solve the following equation:

$$[(12 - 4) \div 2] + [4 \times (5 - 3)] = ?$$

Solution:

$$[(8) \div 2] + [4 \times (2)] = [4] + [8] = 4 + 8 = 12$$

■ EXAMPLE 3.6

Problem: Perform the following operation:

$$-1 + (-1) \times -3$$

Solution: Multiplication and division must be done before addition and subtraction.

$$(-1) \times -3 = 3$$

$$-1 + 3 = 2$$

■ EXAMPLE 3.7

Problem: Perform the following operation:

$$2 \times (-3) + 4$$

Solution: Multiplication and division must be done before addition and subtraction.

$$2 \times (-3) = -6$$

$$-6 + 4 = -2$$

4 Fractions, Decimals, and Percent

Math is ubiquitous; it is invaluable to all aspects of life.

The number 10 divided by 2 gives an exact quotient of 5. This may be written 10/2 = 5. However, if we attempt to divide 7 by 9, we are unable to calculate an exact quotient. This division may be written 7/9 (read "seven ninths"). The number 7/9 represents a number, but not a whole number, and is called a *fraction.* Simply put, fractions are used to express a portion of a whole. The water/waterworks operator is often faced with routine situations that require thinking in fractions and, on occasion, actually working with fractions. One of the common applications for the rules governing the use of fractions in a math problem is dealing with various units of measure. Units such as gallons per minute (gal/min, gpm) are actually fractions. Another example is cubic feet per second (ft^3/sec, cfs). As can be seen, understanding fractions helps in solving a variety of problems.

A fraction is composed of three items: two numbers and a line. The number on the top is the *numerator*, the number on the bottom is the *denominator*, and the line in between them indicates division.

$$\text{Division} \rightarrow \frac{3}{4} \begin{array}{l} \leftarrow \text{Numerator} \\ \leftarrow \text{Denominator} \end{array}$$

The denominator indicates the number of equal-sized pieces into which the entire entity has been cut, and the numerator indicates how many pieces we have.

FRACTIONS

In solving fractions, the following key points are important:

1. Fractions are used to express a portion of a whole.
2. A fraction consists of two numbers separated by a horizontal line or a diagonal line (for example, 1/6).
3. The bottom number, the denominator, indicates the number of equal-sized pieces the whole entity has been cut into.
4. The top number, the numerator, indicates how many pieces we have.
5. Like all other math functions, how we deal with fractions is governed by rules or principles.

Following are some of the principles associated with using fractions:

1. *Same numerator and denominator:* When the numerator and denominator of a fraction are the same, the fraction can be reduced to 1; for example, 5/5 = 1, 33/33 = 1, 69/69 = 1, 34.5/34.5 = 1, and 12/12 = 1.
2. *Whole numbers to fractions:* Any whole number can be expressed as a fraction by placing a "1" in the denominator; for example, 3 is the same as 3/1, and 69 is the same as 69/1.
3. *Adding fractions*: Only fractions with the same denominator can be added, and only the numerators are added. The denominator stays the same. For example, 1/9 + 3/9 = 4/9, and 6/18 + 8/18 = 14/18.
4. *Subtracting fractions:* Only fractions with the same denominator can be subtracted, and only the numerators are subtracted. The denominator remains the same. For example, 7/9 – 4/9 = 3/9 (or 1/3), and 16/30 – 12/30 = 4/30.
5. *Mixed numbers:* A fraction combined with a whole number is a mixed number, such as 4-1/3, 14-2/3, 6-5/7, 43-1/2, and 23-12/35. These numbers are read, respectively, as four and one third; fourteen and two thirds; six and five sevenths; forty-three and one half; and twenty-three and twelve thirty-fifths.
6. *Changing a fraction:* Multiplying the numerator and the denominator by the same number does not change the value of the fraction; for example, 1/3 is the same as (1 × 3)/(3 × 3), which is 3/9.
7. *Simplest terms:* Fractions should be reduced to their simplest terms. This is accomplished by dividing the numerator and denominator by the same number. The result of this division must leave both the numerator and the denominator as whole numbers. For example, 2/6 is not in its simplest terms; by dividing both by 2 we obtain 1/3. The number 2/3 cannot be reduced any further because there is no number that can be divided evenly into the 2 and the 3.

■ Example 4.1

Problem: Reduce the following to their simplest terms:

2/4
14/18
3/4
6/10
9/18
17/29
24/32

Solution:

2/4 = 1/2 (both were divided by 2)
14/18 = 7/9 (both were divided by 2)
3/4 = 3/4 (is already in its simplest form)
6/10 = 3/5 (both were divided by 2)

9/18 = 1/2 (both were divided by 9)
17/29 = 17/29 (is already in its simplest form)
24/32 = 3/4 (both were divided by 8)

8. *Reducing even numbers:* When the starting point is not obvious, do the following: If the numerator and denominator are both even numbers (2, 4, 6, 8, 10, etc.) divide them both by 2, and continue dividing by 2 until a division will no longer yield a whole number for the numerator and the denominator.
9. *Reducing odd numbers:* When the numerator and denominator are both odd numbers (3, 5, 7, 9, 11, 13, 15, 17, etc.), attempt to divide by three, and continue dividing by 3 until a division will no longer yield a whole number with the number and denominator. It is obvious that some numbers such as 5, 7, and 11 cannot be divided by 3 and may in fact be in their simplest terms.
10. *Different denominators:* To add and/or subtract fractions with different denominators, the denominators must be changed to a common denominator (the denominators must be the same). Each fraction must then be converted to a fraction expressing the new denominator. For example, to add 1/8 and 2/5, begin by multiplying the denominators (8 × 5 = 40). Now change 1/8 to a fraction with 40 as the denominator:

    40/8 = 5, 5 × 1 = 5 (the numerator); new fraction is 5/40

    Notice that this is the same as 1/8 except that 5/40 is not reduced to its simplest terms. Change 2/5 to a fraction with 40 as the denominator:

    40/5 = 8, 8 × 2 = 16 (the numerator); new fraction is 16/40

    Complete the addition:

    5/40 + 16/40 = 21/40

11. *Numerator larger:* Any time the numerator is larger than the denominator the fraction should be turned into a mixed number. This is accomplished by the following procedure:
    - Determine the number of times the denominator can be divided evenly into the numerator. This will be the whole number portion of the mixed number.
    - Multiply the whole number times the denominator and subtract from the numerator. This value, the remainder, becomes the numerator of the fraction portion of the mixed number. Consider 28/12:

      28 is divisible by 12 twice, so 2 is the whole number
      2 × 12 = 24
      28/12 – 24/12 = 4/12
      Dividing the top and bottom by 4 gives 1/3
      The new mixed number is 2-1/3

12. *Multiplying fractions:* In order to multiply fractions, simply multiply the numerators and denominators together, and reduce to the simplest terms. For example, find the result of multiplying 1/8 × 2/3:

$$\frac{1}{8}\times\frac{2}{3}=\frac{1\times 2}{8\times 3}=\frac{2}{24}=\frac{1}{12}$$

13. *Dividing fractions:* To divide fractions, simply invert the denominator (turn it upside down), multiply, and reduce to simplest terms; for example, to divide 1/9 by 2/3:

$$\frac{\left(\frac{1}{9}\right)}{\left(\frac{2}{3}\right)}=\frac{1}{9}\times\frac{3}{2}=\frac{1\times 3}{9\times 2}=\frac{3}{18}=\frac{1}{6}$$

Note: The divide symbol can be "÷" or "/" or "—."

14. *Fractions to decimals:* To convert a fraction to a decimal, simply divide the numerator by the denominator; for example,

1/2 = 0.5, 5/8 = 0.625, 7/16 = 0.4375, 1/4 = 0.25

15. *Change inches to feet:* To change inches to feet divide the number of inches by 12; for example,

5/12 = 0.42 feet

■ Example 4.2

Problem: Change the following to feet: 2 inches, 3 inches, 4 inches, 8 inches.

Solution:

2/12 = 0.167 feet
3/12 = 0.25 feet
4/12 = 0.33 feet
8/12 = 0.667 feet

DECIMALS

Although we often use fractions when measuring things, dealing with decimals can be easier when we perform calculations, especially when working with pocket calculators. A decimal is composed of two sets of numbers. The numbers to the left of the decimal are whole numbers, and the numbers to the right of the decimal are parts of a whole number (a fraction of a number), as shown below:

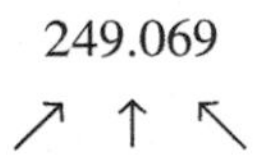

Whole number Decimal Fraction of a number

To solve decimal problems, the following key points are important:

1. We often use fractions when dealing with measurements, but it is usually easier to deal with decimals when we do the calculations, especially when we are working with pocket calculators and computers.
2. We convert a fraction to a decimal by dividing.
3. The horizontal line or diagonal line of the fraction indicates that we divide the bottom number into the top number; for example, to convert 4/5 to a decimal, we divide 4 by 5. Using a pocket calculator, enter the following keystrokes:

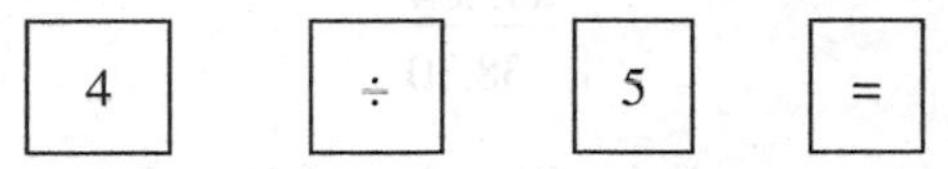

   The display will show the answer: 0.8.
4. Reading numbers: 23,676 is read as twenty-three thousand six hundred seventy-six.
5. Relative values of place:
   - 0.1 tenths
   - 0.01 hundredths
   - 0.001 thousandths
   - 0.0001 ten thousandths
   - 0.00001 hundred thousandths

Note: "And" is not used in reading a whole number but instead is used to signify the presence of a decimal. For example, 23.676 is read as twenty-three and six hundred seventy-six thousandths. 73.2658 is read as seventy-three and two thousand six hundred fifty-eight ten thousandths.

■ Example 4.3

Problem: Reduce the following fractions to decimals:

25/27
2/3
19/64
267/425
32/625
1320/2000

Solution:

25/27 = 0.9259
2/3 = 0.6667
19/64 = 0.2969

267/425 = 0.6282
32/625 = 21.0512
1320/2000 = 62.66

6. When subtracting decimals, simply line up the numbers at the decimal and subtract; for example,

$$\begin{array}{r} 24.66 \\ -13.64 \\ \hline 11.02 \end{array}$$

7. To add decimal numbers, use the same rule as subtraction; in other words, line up the numbers on the decimal and add:

$$\begin{array}{r} 24.66 \\ +13.64 \\ \hline 38.30 \end{array}$$

8. To multiply two or more decimal numbers, follow these basic steps:
   - Multiply the numbers as whole numbers; do not worry about the decimals.
   - Write down the answer.
   - Count the total number of digits (numbers) to the right of the decimal in all of the numbers being multiplied. For example, 3.66 × 8.8 = 32208. There are a total of three numbers to the right of the decimal point (two for the number 3.66 and one for the number 8.8). Therefore, the decimal point would be placed three places to the left from the right side, which results in 32.208.
9. To divide a number by a number containing a decimal, the divisor must be made into a whole number by moving the decimal point to the right until a whole number is obtained.
   - Count the number of places the decimal needs to be moved.
   - Move the decimal in the dividend by the same number of places.

Note: Using a calculator simplifies working with decimals.

PERCENT

The word "percent" means "by the hundred." Percentage is usually designated by the symbol %; thus, 15% means 15 percent or 15/100 or 0.15. These equivalents may be written in the reverse order: 0.15 = 15/100 = 15%. In wastewater treatment, percent is frequently used to express plant performance and for control of biosolids treatment processes. When working with percent, the following key points are important:

- Percents are another way of expressing a part of a whole.
- Percent means "by the hundred," so a percentage is the number out of 100. To determine percent, divide the quantity we wish to express as a percent by the total quantity, then multiply by 100:

$$\text{Percent (\%)} = \frac{\text{Part}}{\text{Whole}} \tag{4.1}$$

For example, 22 percent (or 22%) means 22 out of 100, or 22/100. Dividing 22 by 100 results in the decimal 0.22:

$$22\% = \frac{22}{100} = 0.22$$

- When using percentages in calculations (such as when calculating hypochlorite dosages and when the percent available chlorine must be considered), the percentage must be converted to an equivalent decimal number; this is accomplished by dividing the percentage by 100. For example, calcium hypochlorite (HTH) contains 65% available chlorine. What is the decimal equivalent of 65%? Because 65% means 65 per hundred, divide 65 by 100: 65/100, which is 0.65.
- Decimals and fractions can be converted to percentages. The fraction is first converted to a decimal, then the decimal is multiplied by 100 to get the percentage. For example, if a 50-foot-high water tank has 26 feet of water in it, how full is the tank in terms of the percentage of its capacity?

$$\frac{26\text{ ft}}{50\text{ ft}} = 0.52 \text{ (decimal equivalent)}$$

$$0.52 \times 100 = 52$$

Thus, the tank is 52% full.

■ Example 4.4

Problem: The plant operator removes 6500 gal of biosolids from the settling tank. The biosolids contain 325 gal of solids. What is the percent solids in the biosolids?

Solution:

$$\frac{325\text{ gal}}{6500\text{ gal}} \times 100 = 5\%$$

■ Example 4.5

Problem: Convert 65% to decimal percent.

Solution:

$$\text{Decimal percent} = \frac{\text{Percent}}{100} = \frac{65}{100} = 0.65$$

■ EXAMPLE 4.6

Problem: Biosolids contain 5.8% solids. What is the concentration of solids in decimal percent?

Solution:

$$\text{Decimal percent} = \frac{\text{Percent}}{100} = \frac{5.8}{100} = 0.058$$

Note: Unless otherwise noted, all calculations in the text using percent values require the percent to be converted to a decimal before use.

To determine what quantity a percent equals, first convert the percent to a decimal and then multiply by the total quantity:

$$\text{Quantity} = \text{Total} \times \text{Decimal percent} \tag{4.2}$$

■ EXAMPLE 4.7

Problem: Biosolids drawn from the settling tank are 5% solids. If 2800 gal of biosolids are withdrawn, how many gallons of solids are removed?

Solution:

$$\frac{5}{100} \times 2800 \text{ gal} = 140 \text{ gal}$$

■ EXAMPLE 4.8

Problem: Convert 0.55 to percent.

Solution:

$$0.55 = \frac{55}{100} \times 100 = 55\%$$

(To convert 0.55 to 55%, we multiply by 100, or simply move the decimal point two places to the right.)

■ EXAMPLE 4.9

Problem: Convert 7/22 to a decimal percent to a percent.

Solution:

$$\frac{7}{22} = 0.318; \quad 0.318 \times 100 = 31.8\%$$

■ EXAMPLE 4.10

Problem: Convert the following fractions to decimals:

3/4
3/8
5/8
1/4
7/8
1/2
1/6

Solution:

3/4 = 0.75
3/8 = 0.375
5/8 = 0.625
1/4 = 0.25
7/8 = 0.875
1/2 = 0.5
1/6 = 0.166

Problem: Convert the following percents to decimals:

35%
99%
66%
0.5%
30.4%
22%
88%
3.5%

Solution:

35% = 35/100 = 0.35
99% = 99/100 = 0.99
66% = 66/100 = 0.66
0.5% = 0.5/100 = 0.005
30.4% = 30.4/100 = 0.304
22% = 22/100 = 0.22
88% = 88/100 = 0.88
3.5% = 3.5/100 = 0.035

Problem: Convert the following decimals to percents:

0.66
0.17
0.125
1.0
0.05

Solution:

$0.66 \times 100 = 66\%$
$0.17 \times 100 = 17\%$
$0.125 \times 100 = 12.5\%$
$1.0 \times 100 = 100\%$
$0.05 \times 100 = 5\%$

Problem: Calculate the following ("of" means to multiply; "is" means is equal to):

22% of 110 = ?
17% of 450 = ?
472 is what percentage of 2350?
1.2 is what percentage of 6.3?

Solution:

22% of 110 = $0.22 \times 110 = 24.2$
17% of 450 = $0.17 \times 450 = 76.5$
$472/2350 = 0.2$; $0.2 \times 100 = 20\%$

$$1.2 = x \times 6.3$$

$$\frac{1.2}{6.3} = 0.2 = x$$

$$0.2 \times 100 = 20\%$$

5 Rounding and Significant Digits

The significant figures of a number are those digits that carry meaning, contributing to its precision. However, although the number of significant figures roughly corresponds to precision, it does not correspond to accuracy.

ROUNDING NUMBERS

When rounding numbers, the following key points are important:

1. Numbers are rounded to reduce the number of digits to the right of the decimal point. This is done for convenience, not for accuracy.
2. **Rule:** A number is rounded off by dropping one or more numbers from the right and adding zeros if necessary to place the decimal point. If the last figure dropped is 5 or more, increase the last retained figure by 1. If the last digit dropped is less than 5, do not increase the last retained figure.

■ EXAMPLE 5.1

Problem: Round off 10,546 to 4, 3, 2, and 1 significant figures.

Solution:

10,546 = 10,550 to 4 significant figures
10,546 = 10,500 to 3 significant figures
10,546 = 11,000 to 2 significant figures
10,547 = 10,000 to 1 significant figure

DETERMINING SIGNIFICANT FIGURES

When determining significant figures, the following key points are important:

1. The concept of significant figures is related to rounding.
2. It can be used to determine where to round off.

Note: No answer can be more accurate than the least accurate piece of data used to calculate the answer.

3. **Rule:** Significant figures are those numbers that are known to be reliable. The position of the decimal point does not determine the number of significant figures.

■ **EXAMPLE 5.2**

Problem: How many significant figures are in a measurement of 1.35?

Solution: Three significant figures: 1, 3, and 5.

■ **EXAMPLE 5.3**

Problem: How many significant figures are in a measurement of 0.000135?

Solution: Three significant figures: 1, 3, and 5. The three zeros are used only to place the decimal point.

■ **EXAMPLE 5.4**

Problem: How many significant figures are in a measurement of 103,500?

Solution: Four significant figures: 1, 0, 3, and 5. The remaining two zeros are used to place the decimal point.

■ **EXAMPLE 5.5**

Problem: How many significant figures are in 27,000.0?

Solution: Six significant figures: 2, 7, 0, 0, 0, 0. In this case, the ".0" means that the measurement is precise to 1/10 unit. The zeros indicate measured values and are not used solely to place the decimal point.

6 Powers of Ten and Exponents

$$10^6 = 10 \times 10 \times 10 \times 10 \times 10 \times 10 = 1{,}000{,}000$$

RULES

When working with powers and exponents, the following key points are important:

1. *Powers* are used to identify *area*, as in square feet, and volume, as in *cubic feet*.
2. Powers can also be used to indicate that a number should be squared, cubed, etc. This later designation is the number of times a number must be multiplied by itself. When several numbers are multiplied together, such as $4 \times 5 \times 6 = 120$, the numbers, 4, 5, and 6 are the *factors*; 120 is the *product*.
3. If all of the factors arc alike, such as $4 \times 4 \times 4 \times 4 = 256$, the product is called a *power*. Thus, 256 is a power of 4, and 4 is the *base* of the power. A power is a product obtained by using a base a certain number of times as a factor.
4. Instead of writing $4 \times 4 \times 4 \times 4$, it is more convenient to use an *exponent* to indicate that 4 is to be used as a factor four times. This exponent, a small number placed above and to the right of the base number, indicates how many times the base is to be used as a factor. Using this system of notation, the multiplication $4 \times 4 \times 4 \times 4$ is written as 4^4. The superscript 4 is the exponent, showing that 4 is to be used as a factor 4 times.
5. This same idea can also be applied to letters (*a*, *b*, *x*, *y*, etc.); for example,

$$z^2 = z \times z$$

$$z^4 = z \times z \times z \times z$$

Note: When a number or letter does not have an exponent, it is considered to have an exponent of one.

POWERS

The powers of 1:

$1^0 = 1$
$1^1 = 1$
$1^2 = 1$
$1^3 = 1$
$1^4 = 1$

The powers of 10:

$10^0 = 1$
$10^1 = 10$
$10^2 = 100$
$10^3 = 1000$
$10^4 = 10,000$

EXAMPLES

■ Example 6.1

Problem: How is the term 2^3 written in expanded form?

Solution: The power (exponent) of 3 means that the base number (2) is multiplied by itself three times:

$$2^3 = 2 \times 2 \times 2$$

■ Example 6.2

Problem: How is the term $(3/8)^2$ written in expanded form?

Solution: When parentheses are used, the exponent refers to the entire term within the parentheses. Thus, in this example,

$$(3/8)^2 = 3/8 \times 3/8$$

Note: When a negative exponent is used with a number or term, a number can be re-expressed using a positive exponent:

$$6^{-3} = 1/6^3$$

$$11^{-5} = 1/11^5$$

■ Example 6.3

Problem: How is the term 8^{-3} written in expanded form?

Solution:

$$8^{-3} = 1/8^3 = 1/(8 \times 8 \times 8)$$

Note: A number or letter such as 3^0 or x^0 does not equal 3×1 or $x \times 1$, but simply 1.

7 Averages (Arithmetic Mean) and Median

Whether we speak of harmonic mean, geometric mean, or arithmetic mean, each is designed to find the "center" or "middle" of a set of numbers. These terms capture the intuitive notion of a "central tendency" that may be present in the data. In statistical analysis, an "average of data" is a number that indicates the middle of the distribution of data values. The three most important measures of the center used in statistics are the *mean*, *median*, and *mode*. In this chapter, we discuss the first two.

AVERAGES

An average is a way of representing several different measurements as a single number. Although averages can be useful by indicating about how much or how many, they can also be misleading, as we demonstrate below. You find two kinds of averages in waterworks/wastewater treatment calculations: the *arithmetic mean* (or simply *mean*) and the *median*.

> ***Definition:*** The *mean* (what we usually refer to as an average) is the total of values of a set of observations divided by the number of observations. We simply add up all of the individual measurements and divide by the total number of measurements we took.

■ Example 7.1

Problem: The operator of a waterworks or wastewater treatment plant takes a chlorine residual measurement every day; part of this operating log is shown below. Find the mean.

Day	Chlorine Residual (mg/L)
Monday	0.9
Tuesday	1.0
Wednesday	0.9
Thursday	1.3
Friday	1.1
Saturday	1.4
Sunday	1.2

Solution: Add up the seven chlorine residual readings: 0.9 + 1.0 + 0.9 + 1.3 + 1.1 + 1.4 + 1.2 = 7.8. Next, divide by the number of measurements: 7.8 ÷ 7 = 1.11. The mean chlorine residual for the week was 1.11 mg/L.

MEDIAN

The median is defined as the value of the central item when the data are arrayed by size. First, arrange all of the readings in either ascending or descending order, then find the middle value.

■ EXAMPLE 7.2

Problem: In our chlorine residual example, what is the median?

Solution: Arrange the values in ascending order:

0.9 0.9 1.0 1.1 1.2 1.3 1.4

The middle number is the fourth one, 1.1. So, the median chlorine residual is 1.1 mg/L.

> ***Note:*** Usually the median will be a different value than the mean. If there is an even number of values, you must add one more step, because there is no middle value. You must find the two values in the middle, and then find the mean of those two values.

■ EXAMPLE 7.3

Problem: A water system has four wells with the following capacities: 115 gpm, 100 gpm, 125 gpm, and 90 gpm. What are the mean and the median pumping capacities?

Solution:

$$\text{Mean} = (115 \text{ gpm} + 100 \text{ gpm} + 125 \text{ gpm} + 90 \text{ gpm}) \div 4 = 107.5 \text{ gpm}$$

To find the median, arrange the values in order:

90 gpm 100 gpm 115 gpm 125 gpm

With four values, there is no single middle value, so we must take the mean of the two middle values:

$$\text{Median} = (100 \text{ gpm} + 115 \text{ gpm}) \div 2 = 107.5 \text{ gpm}$$

At times, determining what the original numbers were like is difficult (if not impossible) when dealing only with averages.

■ EXAMPLE 7.4

Problem: A water system has four storage tanks. Three of them have a capacity of 100,000 gal each, while the fourth has a capacity of 1 million gal. What is the mean capacity of the storage tanks?

Solution:

$$(100{,}000 + 100{,}000 + 100{,}000 + 1{,}000{,}000) \div 4 = 325{,}000 \text{ gal}$$

Notice that no tank in this example has a capacity anywhere close to the mean. Determining the median capacity requires us to take the mean of the two middle values; because they are both 100,000 gal, the median is 100,000 gal. Although three of the tanks have the same capacity as the median, this result offers no indication that one of these tanks holds a million gallons, information that could be important for the operator to know.

■ Example 7.5

Problem: Effluent BOD test results for a treatment plant during the month of August are shown below:

Test 1	22 mg/L
Test 2	33 mg/L
Test 3	21 mg/L
Test 4	13 mg/L

What is the average effluent BOD for the month of August?

Solution:

$$(22 \text{ mg/L} + 33 \text{ mg/L} + 21 \text{ mg/L} + 13 \text{ mg/L}) \div 4 = 22.3 \text{ mg/L}$$

■ Example 7.6

Problem: For the primary influent flow, the following composite-sampled solids concentrations were recorded for the week:

Monday	310 mg/L SS
Tuesday	322 mg/L SS
Wednesday	305 mg/L SS
Thursday	326 mg/L SS
Friday	313 mg/L SS
Saturday	310 mg/L SS
Sunday	320 mg/L SS
Total	2206 mg/L SS

What is the average SS?

Solution:

$$\text{Average SS} = \text{Sum of all measurements} \div \text{Number of measurements used}$$

$$= 2206 \text{ mg/L SS} \div 7$$

$$= 315.1 \text{ mg/L SS}$$

8 Solving for the Unknown

MATH RULES OF OPERATION

1. Work from left to right.
2. Do all of the work inside the parentheses first.
3. Do all of the multiplication and division above the line (numerator).
4. Do all of the addition and subtraction above and below the line.
5. Perform the division (divide the numerator by the denominator).

Many water/wastewater calculations involve the use of formulas and equations; for example, process control operations may require the use of equations to solve for an unknown quantity. To make these calculations, you must first know the values for all but one of the terms of the equation to be used. What is an equation? Put simply, an equation is a mathematical statement telling us that what is on one side of an equal (=) sign is equal to what is on the other side; for example, $4 + 5 = 9$. Now, suppose we decide to add 4 to the left side $(4 + 4 + 5)$. What must we do then? We must also add 4 to the right side $(4 + 9)$. Consider the equation $6 + 2 = 8$. If we subtract 3 from the left side $(6 + 2 - 3)$, what must we do next? We must also subtract 3 from the right side $(8 - 3)$. It follows that if the right side were multiplied by a certain number, we must also multiply the left side by that same number. Finally, if one side is divided by, for example, 4, then we must also divide the other side by 4.

The bottom line is that what we do to one side of the equation we must also do to the other side. This is the case, of course, because the two sides, by definition, are always equal.

EQUATIONS

An equation is a statement that two expressions or quantities are equal in value. The statement of equality $6x + 4 = 19$ is an equation; that is, it is algebraic shorthand for "The sum of 6 times a number plus 4 is equal to 19." It can be seen that the equation $6x + 4 = 19$ is much easier to work with than the equivalent sentence. When thinking about equations, it is helpful to consider an equation as being similar to a balance. The equal sign tells you that two quantities are "in balance" (i.e., they are equal). Referring back to the equation $6x + 4 = 19$, it can be solved as follows:

1. $6x + 4 = 19$
2. $6x = 15$
3. $x = 2.5$

Note: Step 1 expresses the whole equation, step 2 shows that 4 has been subtracted from both sides of the equation, and in step 3 both members have been divided by 6.

Note: An equation is kept in balance (both sides of the equal sign are kept equal) by subtracting the same number from both sides (members), adding the same number to both, or dividing or multiplying both by the same number.

The expression $6x + 4 = 19$ is called a *conditional equation* because it is true only when x has a certain value. The number to be found in a conditional equation is called the *unknown number*, the *unknown quantity*, or, more briefly, the *unknown*.

Solving an equation requires finding the value or values of the unknown that make the equation true. Keep in mind that the unknown is a variable that we are trying to solve. The unknown variable is usually represented by a letter, such as x. When solving for an unknown variable, x must be in the numerator or x must be by itself on one side of the equation. If x is the denominator it can trade places with a number on the other side of the = sign. Transpose or flip-flop the equation as shown below. This is the only time you can move x. If x is in the numerator, DO NOT MOVE x.

■ Example 8.1

Problem:

$$\frac{3}{x} = 6$$

Solution:

$$\frac{3}{6} = x$$

$$0.5 = x$$

Let's take a look at another equation:

$$W = F \times D$$

where

W = Work.
F = Force.
D = Distance.

Thus,

$$\text{Work} = \text{Force (lb)} \times \text{Distance (ft or in.)}$$

The resulting quantity is expressed in ft-lb or in.-lb.

Suppose we have this equation:

$$60 = x \times 2$$

How can we determine the value of x? By following the axioms presented below, solving for the unknown is quite simple.

Note: It is important to point out that the following discussion includes only what the axioms are and how they work.

AXIOMS

1. If equal numbers are added to equal numbers, the sums are equal.
2. If equal numbers are subtracted from equal numbers, the remainders are equal.
3. If equal numbers are multiplied by equal numbers, the products are equal.
4. If equal numbers are divided by equal numbers (except zero), the quotients are equal.
5. Numbers that are equal to the same number or to equal numbers are equal to each other.
6. Like powers of equal numbers are equal.
7. Like roots of equal numbers are equal.
8. The whole of anything equals the sum of all its parts.

Note: Axioms 2 and 4 were used to solve the equation $6x + 4 = 19$.

SOLVING EQUATIONS

Solving an equation requires determining the value or values of the unknown number or numbers in the equation.

■ EXAMPLE 8.2

Problem: Find the value of x if $x - 8 = 2$.

Solution: Here it can be seen by inspection that $x = 10$, but inspection does not help in solving more complicated equations. However, notice that to determine that $x = 10$ we can add 8 to each side of the given equation; thus, we have acquired a method or procedure that can be applied to similar but more complex problems.

$$x - 8 = 2$$

Add 8 to each member (axiom 1):

$$x = 2 + 8$$

Collect the terms (that is, add 2 and 8):

$$x = 10$$

■ **Example 8.3**

Problem: Solve for x, if $4x - 4 = 8$ (each side is in simplest terms).

Solution: Add +4 to each side of the equation to get

$$4x = 8 + 4$$

$$4x = 12$$

Now divide both sides:

$$\frac{4x}{4} = \frac{12}{4}$$

$$x = 3$$

■ **Example 8.4**

Problem: Solve for x, if $x + 10 = 15$.

Solution: Subtract 10 from each member (axiom 2):

$$x = 15 - 10$$

Collect the terms:

$$x = 5$$

■ **Example 8.5**

Problem: Solve for x, if $5x + 5 - 7 = 3x + 6$.

Solution: Collect the terms (+5) and (–7):

$$5x - 2 = 3x + 6$$

Add 2 to both members (axiom 2):

$$5x = 3x + 8$$

Subtract $3x$ from both members (axiom 2):

$$2x = 8$$

Divide both members by 2 (axiom 4):

$$x = 4$$

■ Example 8.6

Problem:

$$25 \times x \times 7.48 = 556$$

Solution:

$$x \times (25 \times 7.48) = 556$$
$$x \times 187 = 556$$
$$x = \frac{556}{187}$$
$$x = 2.97$$

■ Example 8.7

Problem:

$$\frac{9x}{3 \times 4} = 24$$

Solution:

$$\frac{9x}{12} = 24$$
$$9x = 24 \times 12$$
$$x = \frac{24 \times 12}{9}$$
$$x = 32$$

■ Example 8.8

Problem:

$$\frac{90}{x} = 4200$$

Solution:

$$\frac{90}{x} = 4200$$
$$\frac{90}{4200} = x$$
$$0.0214 = x$$

■ **Example 8.9**

Problem:

$$9.0 = 3 \times x \times 2.5$$

Solution:

$$9.0 = 7.5 \times x$$

$$\frac{7.5}{9.0} = x$$

$$0.83 = x$$

■ **Example 8.10**

Problem:

$$0.785 \times 0.25 \times 0.25 \times x = 0.53$$

Solution:

$$(0.785 \times 0.25 \times 0.25) \times x = 0.53$$

$$0.0490625x = 0.53$$

$$x = \frac{0.53}{0.0490625}$$

$$x = 10.8$$

■ **Example 8.11**

Problem:

$$\frac{266}{x} = 58$$

Solution:

$$266 = 58 \times x$$

$$\frac{266}{58} = x$$

$$4.59 = x$$

■ **Example 8.12**

Problem:

$$980 = \frac{x}{0.785 \times 80 \times 80}$$

Solution:

$$980 = \frac{x}{5024}$$
$$x = 980 \times 5024$$
$$x = 4{,}923{,}520$$

■ Example 8.13

Problem:

$$x = \frac{155 \times 3 \times 8.34}{0.6}$$

Solution:

$$x = \frac{3878.1}{0.6} = 6463.5$$

■ Example 8.14

Problem:

$$63.5 = \frac{3600}{x \times 8.34}$$

Solution:

$$x = \frac{3600}{63.5 \times 8.34} = \frac{3600}{529.6} = 6.8$$

■ Example 8.15

Problem:

$$116 = \frac{240 \times 1.15 \times 8.34}{0.785 \times 70 \times 70 \times x}$$

Solution:

$$x = \frac{240 \times 1.15 \times 8.34}{0.785 \times 70 \times 70 \times 116} = \frac{2301.2}{446{,}194} = 0.005$$

■ **Example 8.16**

Problem:

$$14 = \frac{x}{188}$$

Solution:

$$x = 14 \times 188 = 2632$$

■ **Example 8.17**

Problem:

$$48 = \frac{100 \times x \times 8.34}{0.785 \times 100 \times 100 \times 6}$$

Solution:

$$48 = \frac{834 \times x}{47{,}100}$$

$$48 \times 47{,}100 = 834 \times x$$

$$x = \frac{48 \times 47{,}100}{834} = 2710.8$$

■ **Example 8.18**

Problem:

$$2.4 = \frac{0.785 \times 6 \times 6 \times 4 \times 7.48}{x}$$

Solution:

$$2.4 \times x = 0.785 \times 6 \times 6 \times 4 \times 7.48$$

$$x = \frac{0.785 \times 6 \times 6 \times 4 \times 7.48}{2.4} = 352.3$$

■ **Example 8.19**

Problem:

$$19{,}878 = 1811.3 \times x$$

Solution:

$$\frac{19{,}878}{1811.3} = x$$

$$10.97 = x$$

■ **Example 8.20**

Problem:

$$\frac{16\times 13\times 1.25\times 7.48}{x}=335$$

Solution:

$$x=\frac{16\times 13\times 1.25\times 7.48}{335}=5.81$$

■ **Example 8.21**

Problem:

$$\frac{x}{4.4\times 8.34}=215$$

Solution:

$$x = 215 \times 4.4 \times 8.34 = 7889.64$$

■ **Example 8.22**

Problem:

$$\frac{x}{248}=2.6$$

Solution:

$$x = 2.6 \times 248 = 644.8$$

■ **Example 8.23**

Problem:

$$8=\frac{x\times 0.16\times 8.34}{64\times 1.2\times 8.34}$$

Solution:

$$\frac{8\times 64\times 1.2\times 8.34}{0.16\times 8.34}=x$$

$$\frac{5124.1}{1.33}=x$$

$$3852.7=x$$

Example 8.24

Problem:

$$\frac{4000 \times 3.5 \times 8.34}{0.785 \times x} = 22.8$$

Solution:

$$\frac{4000 \times 3.5 \times 8.34}{0.785 \times 22.8} = x$$

$$\frac{116,760}{17.9} = x$$

$$6522.9 = x$$

Example 8.25

Problem:

$$111 = \frac{x}{0.785 \times 90 \times 90}$$

Solution:

$$111 \times 0.785 \times 90 \times 90 = x$$

$$705,793.5 = x$$

Example 8.26

Problem:

$$x \times 3.9 \times 8.34 = 3870$$

Solution:

$$x = \frac{3870}{3.9 \times 8.34} = \frac{3870}{32.5} = 119.1$$

Example 8.27

Problem:

$$3.5 = \frac{1,320,000}{x}$$

Solution:

$$x = \frac{1,320,000}{3.5} = 377,142.9$$

■ **Example 8.28**

Problem:

$$0.61 = \frac{160 \times 2.40 \times 8.34}{1970 \times x \times 8.34}$$

Solution:

$$x = \frac{160 \times 2.40 \times 8.34}{1970 \times 0.61 \times 8.34} = \frac{3202.56}{10{,}022.178} = 0.32$$

■ **Example 8.29**

Problem:

$$x^2 \times 0.865 = 3176$$

Note: When solving for x^2 the procedure is the same as solving for x (but there is one extra step at the end). Determine if x^2 is in the numerator, simplify numbers, get x^2 by itself, and solve the equation. Finally, take the square root of both sides.

Solution:

$$x^2 \times 0.865 = 3176$$

$$x^2 = \frac{3176}{0.865}$$

$$x^2 = 3672$$

$$\sqrt{x^2} = \sqrt{3672}$$

$$x = 60.6$$

■ **Example 8.30**

Problem:

$$0.785 \times D^2 = 5038$$

Solution:

$$D^2 = \frac{5038}{0.785}$$

$$D^2 = 6418$$

$$\sqrt{D^2} = \sqrt{6418}$$

$$D = 80.1$$

■ **Example 8.31**

Problem:

$$x^2 \times 12 \times 7.48 = 11{,}112.4$$

Solution:

$$x^2 \times 89.8 = 11{,}112.4$$

$$x^2 = \frac{11{,}112.4}{89.8}$$

$$x^2 = 124$$

$$\sqrt{x^2} = \sqrt{124}$$

$$x = 11.14$$

■ **Example 8.32**

Problem:

$$61 = \frac{75{,}000}{0.785 \times D^2}$$

Solution:

$$D^2 = \frac{75{,}000}{61 \times 0.785}$$

$$D^2 = \frac{75{,}000}{47.9}$$

$$D^2 = 1565.8$$

$$\sqrt{D^2} = \sqrt{1565.8}$$

$$D = 39.57$$

■ **Example 8.33**

Problem:

$$0.785 \times D^2 = 0.66$$

Solution:

$$D^2 = \frac{0.66}{0.785}$$

$$D^2 = 0.8407$$

$$\sqrt{D^2} = \sqrt{0.8407}$$

$$D = 0.917$$

■ **EXAMPLE 8.34**

Problem:

$$2.2 = \frac{0.785 \times D^2 \times 16 \times 7.48}{0.785 \times 90 \times 90}$$

Solution:

$$2.2 = \frac{93.95 \times D^2}{6359}$$

$$2.2 \times 6359 = 93.95 \times D^2$$

$$\frac{2.2 \times 6359}{93.95} = D^2$$

$$148.9 = D^2$$

$$\sqrt{148.9} = \sqrt{D^2}$$

$$12.20 = D$$

CHECKING THE ANSWER

After obtaining a solution to an equation, always check it. This is an easy process. All we need to do is substitute the solution for the unknown quantity in the equation. If the two sides of the equation are identical, the number substituted is the correct answer.

■ **EXAMPLE 8.35**

Problem: Solve and check $4x + 5 - 7 = 2x + 6$.

Solution:

$$4x + 5 - 7 = 2x + 6$$

$$4x - 2 = 2x + 6$$

$$4x = 2x + 8$$

$$2x = 8$$

$$x = 4$$

Substituting the answer $x = 4$ in the original equation,

$$4x + 5 - 7 = 2x + 6$$

$$(4 \times 4) + 5 - 7 = (2 \times 4) + 6$$

$$16 + 5 - 7 = 8 + 6$$

$$14 = 14$$

Because the statement $14 = 14$ is true, the answer $x = 4$ must be correct.

SETTING UP EQUATIONS

The equations discussed to this point were expressed in *algebraic* language. It is important to learn how to set up an equation by translating a sentence into an equation (into algebraic language) and then solving this equation. To set up an equation properly, the following suggestions and examples should help:

1. Always read the statement of the problem carefully.
2. Select the unknown number and represent it by some letter. If more than one unknown quantity exists in the problem, try to represent those numbers in terms of the same letter—that is, in terms of one quantity.
3. Develop the equation using the letter or letters selected and then solve.

■ EXAMPLE 8.36

Problem: One number is eight more than another. The larger number is two less than three times the smaller. What are the two numbers?

Solution: Let n represent the small number. Then $n + 8$ must represent the larger number.

$$n + 8 = 3n - 2$$

$$n = 5 \text{ (small number)}$$

$$n + 8 = 13 \text{ (large number)}$$

■ EXAMPLE 8.37

Problem: If five times the sum of a number and six is increased by three, the result is two less than ten times the number. Find the number.

Solution: Let n represent the number.

$$5(n + 6) + 3 = 10n - 2$$

$$n = 5$$

■ EXAMPLE 8.38

Problem: If $2x + 5 = 10$, solve for x.

Solution:

$$2x + 5 = 10$$

$$2x = 5$$

$$x = 5/2 = 2.5$$

■ **Example 8.39**

Problem: If $0.5x - 1 = -6$, find x.

Solution:

$$0.5x - 1 = -6$$

$$0.5x = -5$$

$$x = -10$$

SUMMARY OF KEY POINTS

- If x is in the numerator, leave x where it is and move the other numbers away from x.
- Only move x if it is in the denominator.
- It does not matter if x is on the left side or the right side of the equation, as $x = 7$ and $7 = x$ mean the same thing.

9 Ratios and Proportions

$$\frac{A}{B} = \frac{C}{D}$$

or

$$A{:}B = C{:}D$$

RATIOS

A ratio is the established relationship between two numbers. For example, if someone says, "I'll give you four to one the Redskins over the Cowboys in the Super Bowl," what does that person mean? Four to one, or 4:1, is a ratio. If someone gives you 4 to 1, it's that person's \$4 against your \$1. As another, more pertinent example, if an average of 3 ft^3 of screenings are removed from each million gallons of wastewater treated, the ratio of screenings removed (ft^3) to treated wastewater (MG) is 3:1. Ratios are normally written using a colon (such as 2:1) or as a fraction (such as 2/1).

PROPORTIONS

A proportion is a statement that two ratios are equal. For example, 1 is to 2 as 3 is to 6, so 1:2 = 3:6. In this case, 1 has the same relationship to 2 that 3 has to 6; in other words, 1 is half the size of 2, and 3 is half the size of 6. Or, alternatively, 2 is twice the size of 1, and 6 is twice the size of 3.

WORKING WITH RATIOS AND PROPORTIONS

When working with ratios and proportions, the following key points are important to remember:

1. One place where fractions are used in calculations is when ratios and proportions are used, such as calculating solutions.
2. A ratio is usually stated in the form "A is to B as C is to D," and we can write it as two fractions that are equal to each other:

$$\frac{A}{B} = \frac{C}{D}$$

3. Cross-multiplying solves ratio problems; that is, we multiply the left numerator (A) by the right denominator (D) and say that the product is equal to the left denominator (B) times the right numerator (C):

$$A \times D = B \times C$$

$$AD = BC$$

4. If one of the four items is unknown, we can solve the ratio by dividing the two known items that are multiplied together by the known item that is multiplied by the unknown. For example, if 2 pounds of alum are needed to treat 500 gallons of water, how many pounds of alum are required to treat 10,000 gallons? We can state this as a ratio: 2 pounds of alum is to 500 gallons of water as x pounds of alum is to 10,000 gallons. This is set up in this manner:

$$\frac{1 \text{ lb alum}}{500 \text{ gal water}} = \frac{x \text{ lb alum}}{10{,}000 \text{ gal water}}$$

Cross-multiplying:

$$500 \times x = 1 \times 10{,}000$$

Transposing:

$$x = \frac{1 \times 10{,}000}{500} = 20 \text{ lb alum}$$

For calculating a proportion, suppose that 5 gallons of fuel cost \$5.40. How much would 15 gallons cost?

$$\frac{5 \text{ gal}}{\$5.40} = \frac{15 \text{ gal}}{x}$$

$$5 \times x = 5.40 \times 15$$

$$x = \frac{5.40 \times 15}{5}$$

$$x = \frac{81}{5}$$

$$x = \$16.20$$

■ Example 9.1

Problem: If a pump will fill a tank in 20 hours at 4 gallons per minute (gpm), how long will it take a 10-gpm pump to fill the same tank?

Solution: First, analyze the problem. Here, the unknown is some number of hours. But should the answer be larger or smaller than 20 hours? If a 4-gpm pump can fill the tank in 20 hours, a larger pump (10-gpm) should be able to complete the filling in less than 20 hours. Therefore, the answer should be less than 20 hours. Now set up the proportion:

$$\frac{x\text{ hr}}{20\text{ hr}} = \frac{4\text{ gpm}}{10\text{ gpm}}$$

$$x = \frac{20 \times 4}{10}$$

$$= 8\text{ hr}$$

■ EXAMPLE 9.2

Problem: Solve for x in the proportion problem given below.

$$\frac{36}{180} = \frac{x}{4450}$$

Solution:

$$\frac{4450 \times 36}{180} = x$$

$$890 = x$$

■ EXAMPLE 9.3

Problem: Solve for the unknown value x in the problem given below.

$$\frac{3.4}{2} = \frac{6}{x}$$

Solution:

$$3.4 \times x = 2 \times 6$$

$$x = \frac{2 \times 6}{3.4} = \frac{12}{3.4} = 3.53$$

■ EXAMPLE 9.4

Problem: One pound of chlorine is dissolved in 65 gallons of water. To maintain the same concentration, how many pounds of chlorine would have to be dissolved in 150 gallons of water?

Solution:

$$\frac{1\text{ lb}}{65\text{ gal}} = \frac{x\text{ lb}}{150\text{ gal}}$$

$$65\text{ gal} \times x = 1 \times 150$$

$$x = \frac{1 \times 150}{65}$$

$$x = 2.3\text{ lb}$$

■ Example 9.5

Problem: It takes 5 workers 50 hours to complete a job. At the same rate, how many hours would it take 8 workers to complete the job?

Solution:

$$\frac{5\text{ workers}}{8\text{ workers}} = \frac{x\text{ hr}}{50\text{ hr}}$$

$$5 \times 50 = 8 \times x$$

$$\frac{5 \times 50}{8} = x$$

$$31.3\text{ hr} = x$$

10 Electrical Calculations

Basic electrical conversion example problems and solutions were presented in Chapter 2. This chapter provides a more in-depth discussion of typical direct current (DC) electrical calculations because many state licensure examinations include basic, practical electrical problems. Alternating current (AC) is not presented here because, for the water/wastewater operator, electrical knowledge beyond Ohm's law and basic circuit analysis is not required. Only the very basics are presented here, as operators who take these exams are required to perform only very basic electrical computations. Moreover, licensure aside, many operators are required to perform basic electrical maintenance and installations that may involve basic computations. It should be pointed out, however, that only qualified electricians should perform electrical work of any type.

ELECTRICAL CIRCUITS AND POWER

An electrical circuit includes

- An energy source—a source of electromotive force (emf) or voltage, such as a battery or generator
- A conductor (wire)
- A load
- A means of control

The energy source could be a battery, as shown in Figure 10.1, or some other means of producing a voltage. The load that dissipates the energy could be a lamp, a resistor, or some other device that does useful work, such as an electric toaster, a power drill, a radio, or a soldering iron. Conductors are wires that offer low resistance to current; they connect all of the loads in the circuit to the voltage source. No electrical device dissipates energy unless current flows through it. Because conductors, or wires, are not perfect conductors they heat up (dissipate energy), so they are actually part of the load. For simplicity, however, we usually think of the connecting wiring as having no resistance, as it would be tedious to assign a very low resistance value to the wires every time we wanted to solve a problem. Control devices might be switches, variable resistors, circuit breakers, fuses, or relays.

OHM'S LAW

Simply put, Ohm's law defines the relationship between current, voltage, and resistance in electric circuits. Ohm's law can be expressed mathematically in three ways:

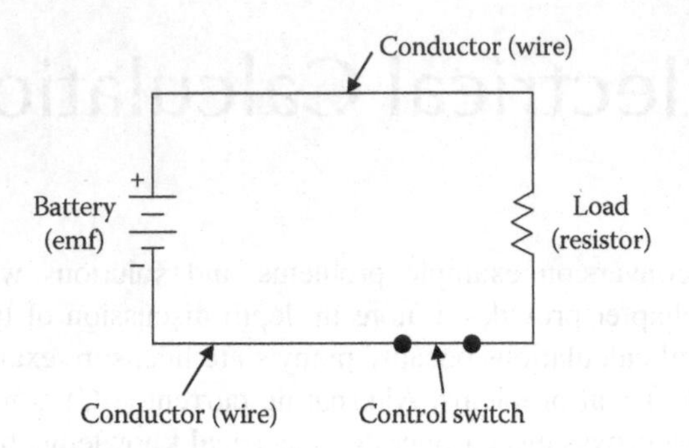

FIGURE 10.1 Simple closed circuit.

1. The *current* (I) in a circuit is equal to the voltage applied to the circuit divided by the resistance of the circuit. Stated another way, the current in a circuit is *directly* proportional to the applied voltage and *inversely* proportional to the circuit resistance. Ohm's law may be expressed as

$$I = \frac{E}{R} \tag{10.1}$$

   where

   I = Current in amps.
   E = Voltage in volts.
   R = Resistance in ohms.

2. The *resistance* (R) of a circuit is equal to the voltage applied to the circuit divided by the current in the circuit:

$$R = \frac{E}{I} \tag{10.2}$$

3. The applied *voltage* (E) to a circuit is equal to the product of the current and the resistance of the circuit:

$$E = I \times R \tag{10.3}$$

If any two of the quantities in Equations 10.1 through 10.3 are known, the third may be easily found. Let's look at an example.

■ Example 10.1

Problem: Figure 10.2 shows a circuit containing a resistance (R) of 6 ohms and a source of voltage (E) of 3 volts. How much current (I) flows in the circuit?
Solution:

$$I = \frac{E}{R} = \frac{3}{6} = 0.5 \text{ amperes}$$

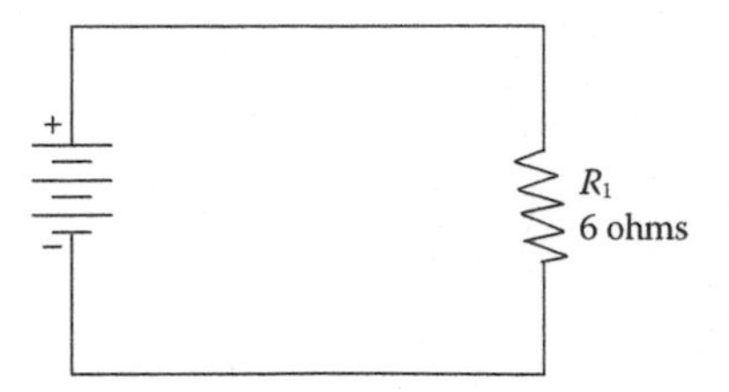

FIGURE 10.2 Determining current in a simple circuit.

To observe the effect of source voltage on circuit current, in the next example we use the circuit shown in Figure 10.2 but double the voltage to 6 volts.

■ Example 10.2

Problem: Given that $E = 6$ volts and $R = 6$ ohms, what is I?

Solution:

$$I = \frac{E}{R} = \frac{6}{6} = 1 \text{ ampere}$$

Notice that as the source of voltage doubles, the circuit current also doubles.

> ***Note:*** Circuit current is directly proportional to applied voltage and will change by the same factor that the voltage changes.

To verify that current is inversely proportional to resistance, assume that the resistor in Figure 10.2 has a value of 12 ohms.

■ Example 10.3

Problem: Given that $E = 3$ volts and $R = 12$ ohms, what is I?

Solution:

$$I = \frac{E}{R} = \frac{3}{12} = 0.25 \text{ ampere}$$

Comparing the current of 0.25 amp for the 12-ohm resistor to the 0.5-amp current obtained with the 6-ohm resistor shows that doubling the resistance will reduce the current to one-half the original value. The point here is that *circuit current is inversely proportional to the circuit resistance.*

Recall that, if we know any two quantities (E, I, or R), we can calculate the third. In many circuit applications, current is known and either the voltage or the resistance will be the unknown quantity. To solve a problem in which voltage (E) and resistance (R) are known, the basic formula for Ohm's law must be transposed to solve for I. The Ohm's law equations can be memorized and practiced effectively by using an Ohm's law circle (see Figure 10.3). To find the equation for E, I, or R when two quantities are known, cover the unknown third quantity with your finger, ruler, or piece of paper as shown in Figure 10.4.

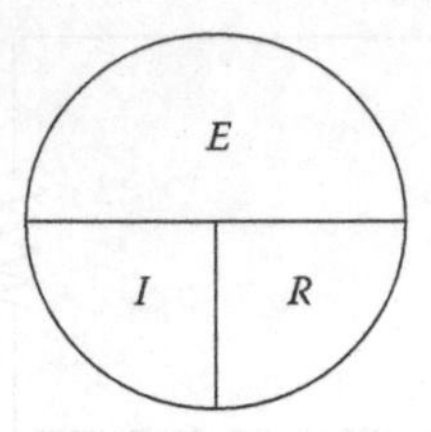

FIGURE 10.3 Ohm's law circle.

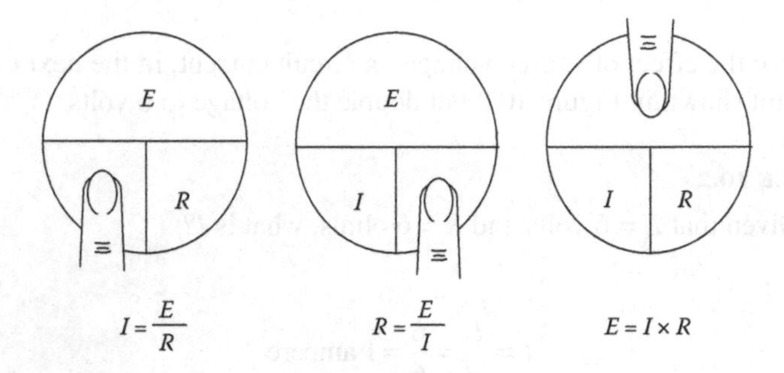

FIGURE 10.4 Putting Ohm's law circle to work.

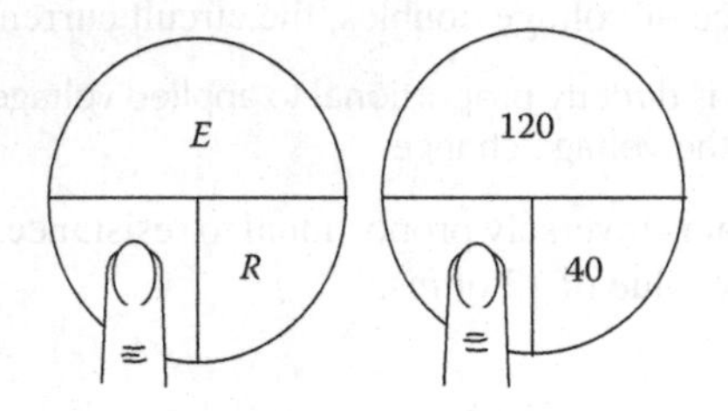

FIGURE 10.5 Ohm's law illustration for Example 10.4.

■ Example 10.4

Problem: Find I when E = 120 V and R = 40 W.

Solution: Place finger on I as shown in Figure 10.5. Use Equation 10.1 to find the unknown I:

$$I = \frac{E}{R} = \frac{120}{40} = 3 \text{ amperes}$$

■ Example 10.5

Problem: Find R when E = 220 V and I = 10 A.

Solution: Place finger on R as shown in Figure 10.6. Use Equation 10.2 to find the unknown R:

$$R = \frac{E}{I} = \frac{220}{10} = 22 \text{ ohms}$$

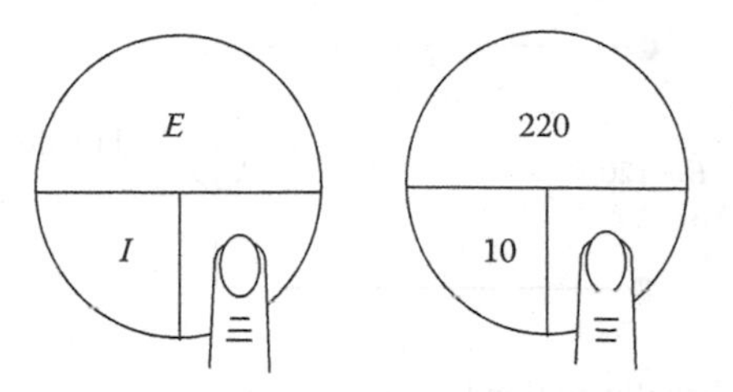

FIGURE 10.6 Ohm's law illustration for Example 10.5.

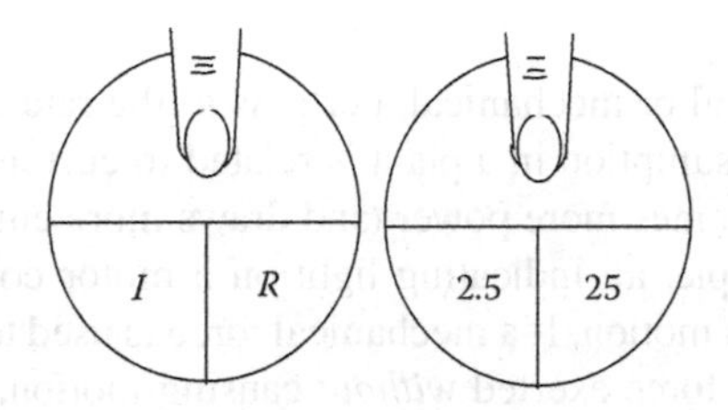

FIGURE 10.7 Ohm's law illustration for Example 10.6.

■ Example 10.6

Problem: Find E when $I = 2.5$ A and $R = 25$ W.

Solution: Place finger on E as shown in Figure 10.7. Use Equation 10.3 to find the unknown E:

$$E = I \times R = 2.5 \times 25 = 62.5 \text{ V}$$

> ***Note:*** In the previous examples, we have demonstrated how the Ohm's law circle can help solve simple voltage, current, and amperage problems. Beginning students are cautioned, however, not to rely entirely on the use of this circle when transposing simple formulas but rather to use it to supplement their knowledge of the algebraic method. Algebra is a basic tool in the solution of electrical problems. The importance of knowing how to use it should not be underemphasized, and its use should not be bypassed after the operator has learned a shortcut method such as the one indicated in this circle.

■ Example 10.7

Problem: An electric light bulb draws 0.5 A when operating on a 120-V DC circuit. What is the resistance of the bulb?

Solution: The first step in solving a circuit problem is to sketch a schematic diagram of the circuit itself, labeling each of the parts and showing the known values (see Figure 10.8). Because I and E are known, we can use Equation 10.2 to solve for R:

$$R = \frac{E}{I} = \frac{120}{0.5} = 240 \text{ ohms}$$

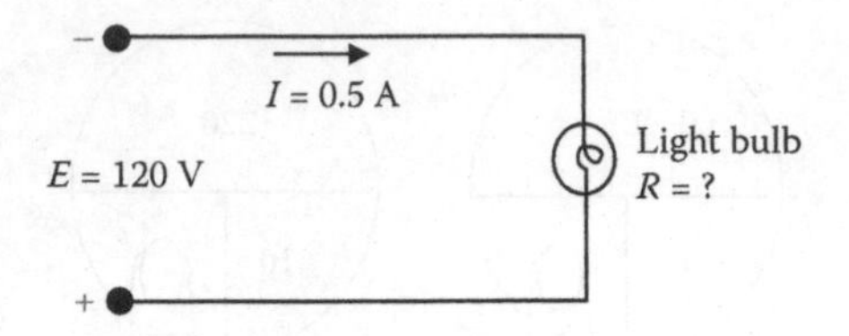

FIGURE 10.8 Ohm's law illustration for Example 10.7.

Electrical Power

Power, whether electrical or mechanical, pertains to the rate at which work is being done, so the power consumption in a plant is related to current flow. A large electric motor or air dryer consumes more power (and draws more current) in a given length of time than, for example, an indicating light on a motor controller. Work is done whenever a force causes motion. If a mechanical force is used to lift or move a weight, work is done; however, force exerted *without* causing motion, such as the force of a compressed spring acting between two fixed objects, does not constitute work.

Note: Power is the rate at which work is done.

Electrical Power Calculations

The electrical power (P) used in any part of a circuit is equal to the voltage (E) across that part of the circuit multiplied by the current (I) in that part. In equation form:

$$P = E \times I \tag{10.4}$$

where

P = Power (watts, W).
E = Voltage (volts, V).
I = Current (amps, A).

If we know the current (I) and the resistance (R) but not the voltage, we can find the power (P) by using Ohm's law for voltage, so by substituting Equation 10.3:

$$E = I \times R$$

into Equation 10.4, we obtain:

$$P = (I \times R) \times I = I^2 \times R \tag{10.5}$$

In the same manner, if we know the voltage and the resistance but not the current, we can find the P by using Ohm's law for current, so by substituting Equation 10.1:

$$I = \frac{E}{R}$$

into Equation 10.4, we obtain:

$$P = E \times \frac{E}{R} = \frac{E^2}{R} \tag{10.6}$$

Note: If we know any two quantities, we can calculate the third.

■ Example 10.8

Problem: The current through a 200-W resistor to be used in a circuit is 0.25 A. Find the power rating of the resistor.

Solution: Because the current (*I*) and resistance (*R*) are known, we can use Equation 10.5 to find *P*:

$$P = I^2 \times R = (0.25)^2 \times 200 = 0.0625 \times 200 = 12.5 \text{ W}$$

■ Example 10.9

Problem: How many kilowatts of power are delivered to a circuit by a 220-V generator that supplies 30 A to the circuit?

Solution: Because the voltage (*E*) and current (I) are given, we can use Equation 10.4 to find *P*:

$$P = E \times I = 220 \times 30 = 6600 \text{ W} = 6.6 \text{ kW}$$

■ Example 10.10

Problem: If the voltage across a 30,000-W resistor is 450 V, what is the power dissipated in the resistor?

Solution: Because the resistance (*R*) and voltage (*E*) are known, we can use Equation 10.6 to find *P*:

$$P = \frac{E^2}{R} = \frac{(450)^2}{30{,}000} = \frac{202{,}500}{30{,}000} = 6.75 \text{ watts}$$

In this section, *P* was expressed in terms of various pairs of the other three basic quantities *E*, *I*, and *R*. In practice, we should be able to express any one of the three basic quantities, as well as *P*, in terms of any two of the others. Figure 10.9 is a summary of the 12 basic formulas we should know. The four quantities *E*, *I*, *R*, and *P* are at the center of the figure. Adjacent to each quantity are three segments. Note that in each segment the basic quantity is expressed in terms of two other basic quantities, and no two segments are alike.

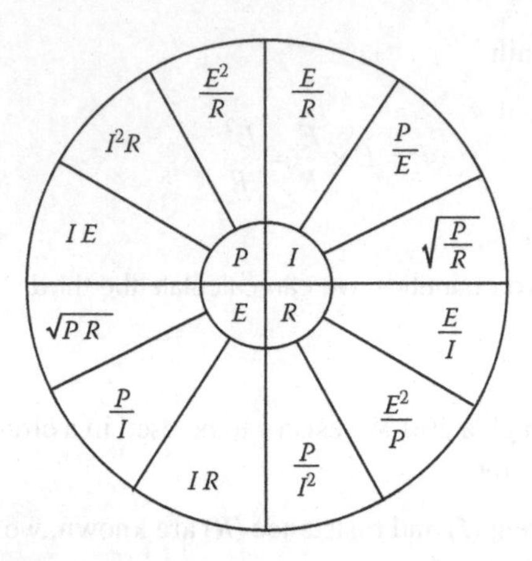

FIGURE 10.9 Ohm's law circle—summary of basic formulas.

Electrical Energy

Energy can be defined as the ability to do work (energy and time are essentially the same and are expressed in identical units). Energy is expended when work is done, because it takes energy to maintain a force when that force acts through a distance. The total energy expended to do a certain amount of work is equal to the working force multiplied by the distance through which the force moves to do the work. In electricity, total energy expended is equal to the *rate* at which work is done, multiplied by the length of time the rate is measured. Essentially, energy (W) is equal to power (P) times time (t). The kilowatt-hour (kWh) is a unit commonly used for large amounts of electric energy or work. The amount of kilowatt-hours is calculated as the product of the power in kilowatts (kW) and the time in hours (hr) during which the power is used:

$$\text{kWh} = \text{kW} \times \text{hr} \tag{10.7}$$

■ Example 10.11

Problem: How much energy is delivered in 4 hours by a generator supplying 12 kW?

Solution:

$$\text{kWh} = \text{kW} \times \text{hr} = 12 \times 4 = 48 \text{ kWh}$$

SERIES DC CIRCUIT CHARACTERISTICS

An electrical circuit is made up of a voltage source, the necessary connecting conductors, and the effective load. If the circuit is arranged so the electrons have only *one* possible path, the circuit is a *series circuit*. A series circuit, then, is defined as a circuit that contains only one path for current flow. Figure 10.10 shows a series circuit having several loads (resistors).

Note: A series circuit is a circuit having only one path for the current to flow along.

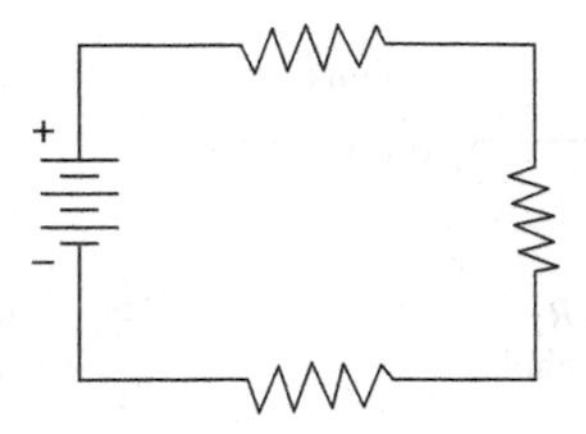

FIGURE 10.10 Series circuit.

Series Circuit Resistance

To follow its electrical path, the current in a series circuit must flow through resistors inserted in the circuit (see Figure 10.10); thus, each additional resistor offers added resistance. In a series circuit, the total circuit resistance (R_T) is equal to the sum of the individual resistances, or

$$R_T = R_1 + R_2 + R_3 + \ldots + R_n \tag{10.8}$$

where

R_T = Total resistance (W).
R_1, R_2, R_3 = Resistance in series (W).
R_n = Any number of additional resistors in the series.

■ Example 10.12

Problem: Three resistors of 10 ohms, 12 ohms, and 25 ohms are connected in series across a battery whose emf is 110 volts (Figure 10.11). What is the total resistance?

Solution:

Given:

R_1 = 10 ohms
R_2 = 12 ohms
R_3 = 25 ohms

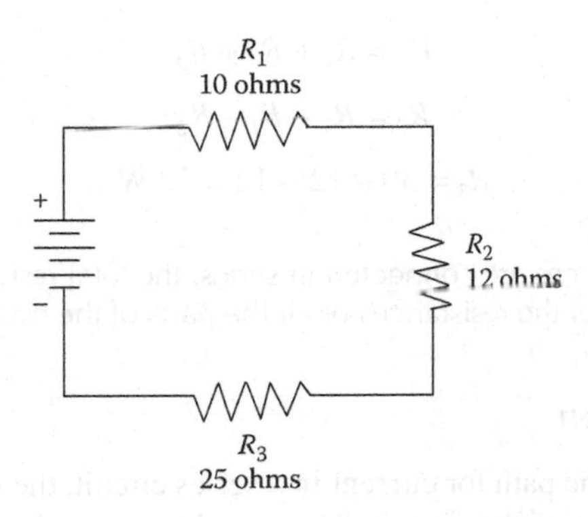

FIGURE 10.11 Solving for total resistance in a series circuit in Example 10.12.

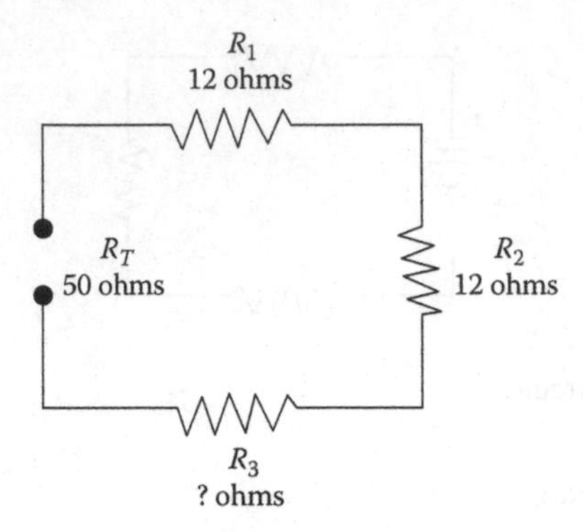

FIGURE 10.12 Calculating the value of one resistance in a series circuit in Example 10.13.

$$R_T = R_1 + R_2 + R_3$$

$$R_T = 10 + 12 + 25 = 47 \text{ W}$$

Equation 10.8 can be transposed to solve for the value of an unknown resistance; for example, transposition can be used in some circuit applications where the total resistance is known but the value of a circuit resistor has to be determined.

■ Example 10.13

Problem: The total resistance of a circuit containing three resistors is 50 ohms (see Figure 10.12). Two of the circuit resistors are 12 ohms each. Calculate the value of the third resistor (R_3).

Solution:

Given:

$R_T = 50$ ohms
$R_1 = 12$ ohms
$R_2 = 12$ ohms

$$R_T = R_1 + R_2 + R_3$$

$$R_3 = R_T - R_1 - R_2$$

$$R_3 = 50 - 12 - 12 = 26 \text{ W}$$

Note: When resistances are connected in series, the total resistance in the circuit is equal to the sum of the resistances of all the parts of the circuit.

Series Circuit Current

Because there is but one path for current in a series circuit, the same current (I) must flow through each part of the circuit. Thus, to determine the current throughout a series circuit, only the current through one of the parts must be known. The fact that

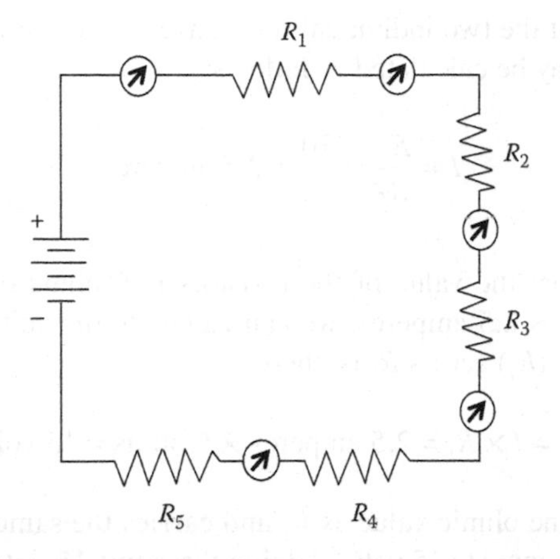

FIGURE 10.13 Current in a series circuit.

the same current flows through each part of a series circuit can be verified by inserting ammeters into the circuit at various points as shown in Figure 10.13, where each meter indicates the same value of current.

> ***Note:*** In a series circuit, the same current flows in every part of the circuit. Do not add the currents in each part of the circuit to obtain *I*.

Series Circuit Voltage

The *voltage drop* across the resistor in the basic circuit is the total voltage across the circuit and is equal to the applied voltage. The total voltage across a series circuit is also equal to the applied voltage but consists of the sum of two or more individual voltage drops. This statement can be proven by an examination of the circuit shown in Figure 10.14. In this circuit, a source potential (E_T) of 30 volts is impressed across a series circuit consisting of two 6-ohm resistors. The total resistance of the circuit

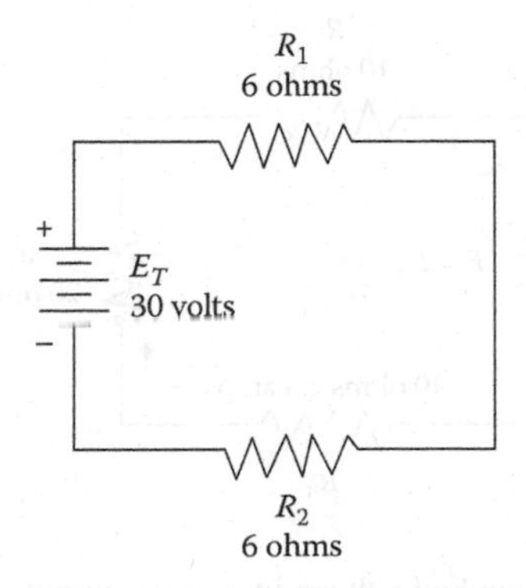

FIGURE 10.14 Calculating total resistance in a series circuit.

is equal to the sum of the two individual resistances, or 12 ohms. Using Ohm's law, the circuit current may be calculated as follows:

$$I = \frac{E_T}{R_T} = \frac{30}{12} = 2.5 \text{ amperes}$$

Because we know that the value of the resistors is 6 ohms each, and the current through the resistors is 2.5 amperes, we can calculate the voltage drops across the resistors. The voltage (E_1) across R_1 is, therefore,

$$E_1 = I \times R_1 = 2.5 \text{ amperes} \times 6 \text{ ohms} = 15 \text{ volts}$$

Because R_2 is the same ohmic value as R_1 and carries the same current, the voltage drop across R_2 is also equal to 15 volts. Adding these two 15-volt drops together gives a total drop of 30 volts, exactly equal to the applied voltage. For a series circuit then,

$$E_T = E_1 + E_2 + E_3 + \ldots + E_n \tag{10.9}$$

where

E_T = Total voltage (V).
E_1 = Voltage across resistance R_1 (V).
E_2 = Voltage across resistance R_2 (V).
E_3 = Voltage across resistance R_3 (V).

■ Example 10.14

Problem: A series circuit consists of three resistors having values of 10 ohms, 20 ohms, and 40 ohms. Find the applied voltage if the current through the 20-ohm resistor is 2.5 amp.

Solution: To solve this problem, first draw a circuit diagram and label it as shown in Figure 10.15.

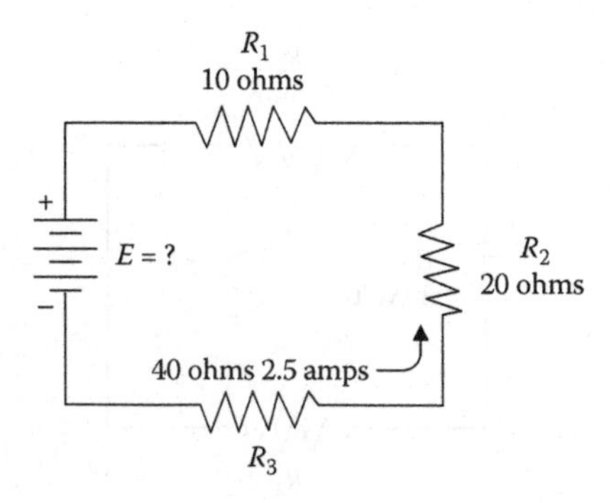

FIGURE 10.15 Solving for applied voltage in a series circuit in Example 10.14.

Given:

$R_1 = 10$ ohms
$R_2 = 20$ ohms
$R_3 = 40$ ohms
$I = 2.5$ amperes

Because the circuit involved is a series circuit, the same 2.5 amperes of current flow through each resistor. Using Ohm's law, the voltage drops across each of the three resistors can be calculated:

$$E_1 = 25 \text{ volts}$$

$$E_2 = 50 \text{ volts}$$

$$E_3 = 100 \text{ volts}$$

When the individual drops are known, they can be added to find the total or applied voltage by using Equation 10.9:

$$E_T = E_1 + E_2 + E_3$$

$$E_T = 25 \text{ V} + 50 \text{ V} + 100 \text{ V} = 175 \text{ V}$$

Note: The total voltage (E_T) across a series circuit is equal to the sum of the voltages across each resistance of the circuit.

Note: The voltage drops that occur in a series circuit are in direct proportion to the resistance across which they appear. This is the result of having the same current flow through each resistor. Thus, the larger the resistor, the larger will be the voltage drop across it.

Series Circuit Power

Each resistor in a series circuit consumes power. This power is dissipated in the form of heat. Because this power must come from the source, the total power must be equal in amount to the power consumed by the circuit resistances. In a series circuit, the total power is equal to the sum of the powers dissipated by the individual resistors. Total power (P_T) is thus equal to

$$P_T = P_1 + P_2 + P_3 + \ldots + P_n \tag{10.10}$$

where

P_T = Total power (W).
P_1 = Power used in first part (W).
P_2 = Power used in second part (W).
P_3 = Power used in third part (W).
P_n = Power used in *n*th part (W).

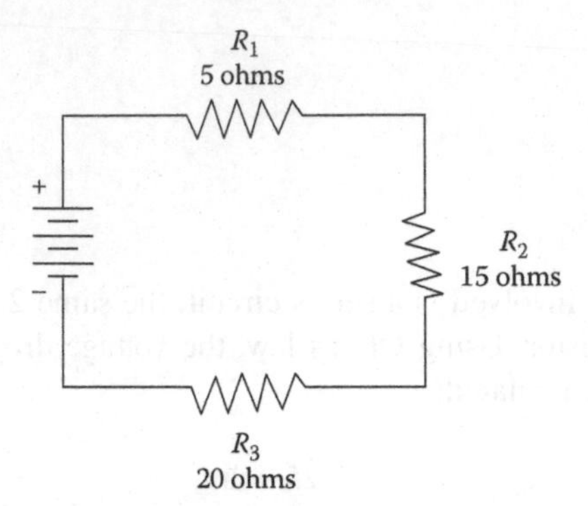

FIGURE 10.16 Solving for total power in a series circuit in Example 10.16.

■ Example 10.15

Problem: A series circuit consists of three resistors having values of 5 ohms, 15 ohms, and 20 ohms. Find the total power dissipation when 120 volts is applied to the circuit (see Figure 10.16).

Solution:

Given:

$R_1 = 5$ ohms
$R_2 = 15$ ohms
$R_3 = 20$ ohms
$E = 120$ volts

The total resistance is found first:

$$R_T = R_1 + R_2 + R_3 = 5 \text{ ohms} + 15 \text{ ohms} + 20 \text{ ohms} = 40 \text{ ohms}$$

Using total resistance and the applied voltage, we can calculate the circuit current:

$$I = \frac{E_T}{R_T} = \frac{120}{40} = 3 \text{ amperes}$$

Using the power formula, we can calculate the individual power dissipations:

For resistor R_1:

$$P_1 = I^2 \times R_1 = (3)^2 \times 5 = 45 \text{ watts}$$

For resistor R_2:

$$P_2 = I^2 \times R_2 = (3)^2 \times 15 = 135 \text{ watts}$$

For resistor R_3:

$$P_3 = I^2 \times R_3 = (3)^2 \times 20 = 180 \text{ watts}$$

To obtain total power:

$$P_T = P_1 + P_2 + P_3 = 45 \text{ watts} + 135 \text{ watts} + 180 \text{ watts} = 360 \text{ watts}$$

To check our answer, the total power delivered by the source can be calculated:

$$P = E \times I = 120 \text{ volts} \times 3 \text{ amperes} = 360 \text{ watts}$$

Thus, the total power is equal to the sum of the individual power dissipations.

> ***Note:*** We found that Ohm's law can be used for total values in a series circuit as well as for individual parts of the circuit. Similarly, the formula for power may be used for total values:
>
> $$P_T = E_T \times I \tag{10.11}$$

General Series Circuit Analysis

Now that we have discussed the pieces involved in solving the puzzle that is series circuit analysis, we can move on to the next step in the process: solving series circuit analysis in total.

■ Example 10.16

Problem: Three resistors of 20 ohms, 20 ohms, and 30 ohms are connected across a battery supply rated at 100-volt terminal voltage. Completely solve the circuit shown in Figure 10.17.

> ***Note:*** To solve the circuit, the total resistance must be found first, then the circuit current can be calculated. When the current is known, the voltage drops and power dissipations can be calculated.

Solution: The total resistance is

$$R_T = R_1 + R_2 + R_3 = 20 \text{ ohms} + 20 \text{ ohms} + 30 \text{ ohms} = 70 \text{ ohms}$$

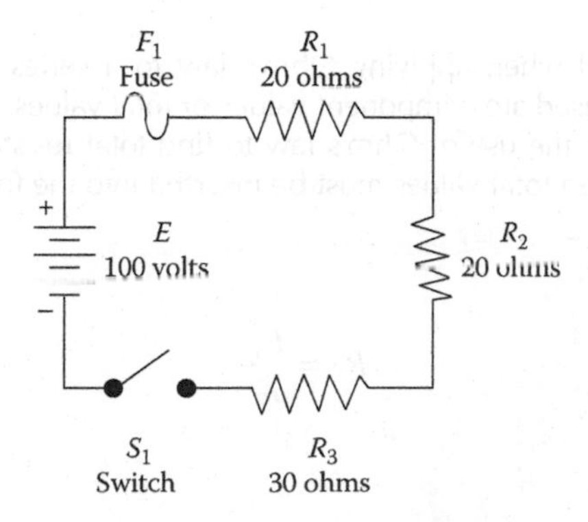

FIGURE 10.17 Solving for various values in a series circuit in Example 10.16.

By Ohm's law, the current is

$$I = \frac{E}{R_T} = \frac{100}{70} = 1.43 \text{ amperes}$$

The voltage (E_1) across R_1 is

$$E_1 = I \times R_1 = 1.43 \text{ amperes} \times 20 \text{ ohms} = 28.6 \text{ volts}$$

The voltage (E_2) across R_2 is

$$E_2 = I \times R_2 = 1.43 \text{ amperes} \times 20 \text{ ohms} = 28.6 \text{ volts}$$

The voltage (E_3) across R_3 is

$$E_3 = I \times R_2 = 1.43 \text{ amperes} \times 30 \text{ ohms} = 42.9 \text{ volts}$$

The power dissipated by R_1 is

$$P_1 = E_1 \times I = 28.6 \text{ volts} \times 1.43 \text{ amperes} = 40.9 \text{ watts}$$

The power dissipated by R_2 is

$$P_2 = E_2 \times I = 28.6 \text{ volts} \times 1.43 \text{ amperes} = 40.9 \text{ watts}$$

The power dissipated by R_3 is

$$P_3 = E_3 \times I = 42.9 \text{ volts} \times 1.43 \text{ amperes} = 61.3 \text{ watts}$$

The total power dissipated is

$$P_T = E_T \times I = 100 \text{ volts} \times 1.43 \text{ amperes} = 143 \text{ watts}$$

Note: Keep in mind when applying Ohm's law to a series circuit to consider whether the values used are component values or total values. When the information available allows the use of Ohm's law to find total resistance, total voltage, and total current, then total values must be inserted into the formula.

To find total resistance:

$$R_T = \frac{E_T}{I_T}$$

To find total voltage:

$$E_T = I_T \times R_T$$

To find total current:

$$I_T = \frac{E_T}{R_T}$$

PARALLEL DC CIRCUITS

The principles we applied to solving simple series circuit calculations for determining the reactions of such quantities as voltage, current, and resistance can be used in parallel and series–parallel circuits.

Parallel Circuit Characteristics

A parallel circuit is defined as one having two or more components connected across the same voltage source (see Figure 10.18). Recall that a series circuit has only one path for current flow. As additional loads (resistors, etc.) are added to the circuit, the total resistance increases and the total current decreases. This is *not* the case in a parallel circuit. In a parallel circuit, each load (or branch) is connected directly across the voltage source. In Figure 10.18, commencing at the voltage source (E_b) and tracing counterclockwise around the circuit, two complete and separate paths can be identified in which current can flow. One path is traced from the source through resistance R_1 and back to the source, the other from the source through resistance R_2 and back to the source.

Voltage in Parallel Circuits

Recall that in a series circuit the source voltage divides proportionately across each resistor in the circuit. In a parallel circuit (see Figure 10.18), the same voltage is present across all of the resistors of a parallel group. This voltage is equal to the applied voltage (E_b) and can be expressed in equation form as

$$E_b = E_{R1} = E_{R2} = E_{Rn}$$

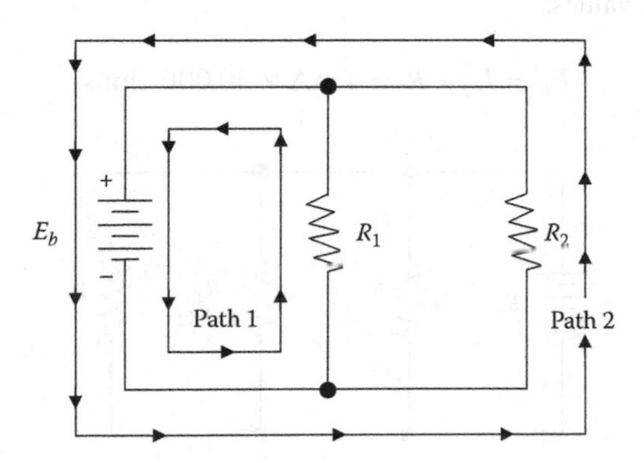

FIGURE 10.18 Basic parallel circuit.

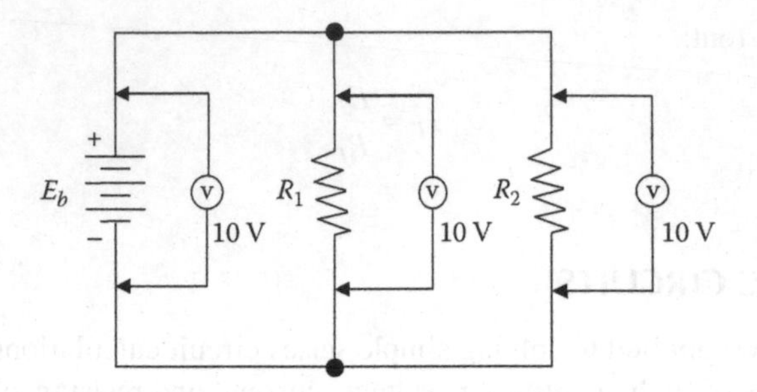

FIGURE 10.19 Voltage comparison in a parallel circuit.

We can verify Equation 10.11 by taking voltage measurements across the resistors of a parallel circuit, as illustrated in Figure 10.19. Notice that each voltmeter indicates the same amount of voltage; that is, the voltage across each resistor is the same as the applied voltage.

Note: In a parallel circuit, the voltage remains the same throughout the circuit.

■ Example 10.17

Problem: Assume that the current through a resistor of a parallel circuit is known to be 4 milliamperes (mA) and the value of the resistor is 40,000 ohms. Determine the potential (voltage) across the resistor. The circuit is shown in Figure 10.20.

Solution:

Given:

$R_2 = 40{,}000$ ohms

$I_{R2} = 4$ mA

Find E_{R2} and E_b. Select the appropriate equation:

$$E = I \times R$$

Substitute known values:

$$E_{R2} = I_{R2} \times R_2 = 4 \text{ mA} \times 40{,}000 \text{ ohms}$$

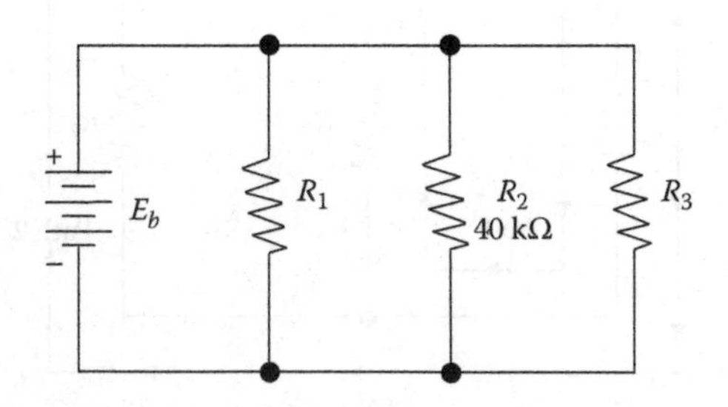

FIGURE 10.20 Illustration for Example 10.17.

Using power of tens,

$$E_{R2} = (4 \times 10^{-3}) \times (40 \times 10^{3}) = 4.0 \times 40 = 160 \text{ volts}$$

Therefore,

$$E_b = 160 \text{ volts}$$

Current in Parallel Circuits

In a series circuit, a single current flows. Its value is determined in part by the total resistance of the circuit; however, the source current in a parallel circuit divides among the available paths in relation to the value of the resistors in the circuit. Ohm's law remains unchanged. For a given voltage, current varies inversely with resistance.

> ***Note:*** Ohm's law states that the *current in a circuit is inversely proportional to the circuit resistance*. This fact, important as a basic building block of electrical theory, is also important in the following explanation of current flow in parallel circuits.

The behavior of current in a parallel circuit is best illustrated by example (see Figure 10.21). The resistors R_1, R_2, and R_3 are in parallel with each other and with the battery. Each parallel path is then a branch with its own individual current. When the total current (I_T) leaves the voltage source (E), part I_1 of current I_T will flow through R_1, part I_2 will flow through R_2, and I_3 through R_3. The branch currents I_1, I_2, and I_3 can be different; however, if a voltmeter (used for measuring the voltage of a circuit) is connected across R_1, R_2, and R_3, then the respective voltages E_1, E_2, and E_3 will be equal. Therefore,

$$E = E_1 = E_2 = E_3 \tag{10.12}$$

The total current, I_T, is equal to the sum of all branch currents:

$$I_T = I_1 + I_2 + I_3 \tag{10.13}$$

This formula applies for any number of parallel branches, whether the resistances are equal or unequal.

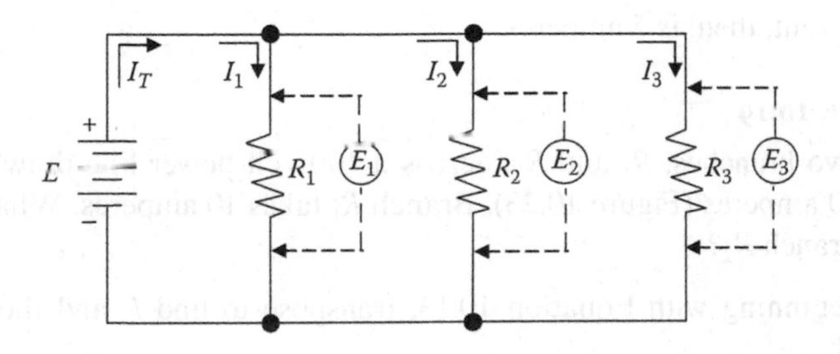

FIGURE 10.21 Parallel circuit.

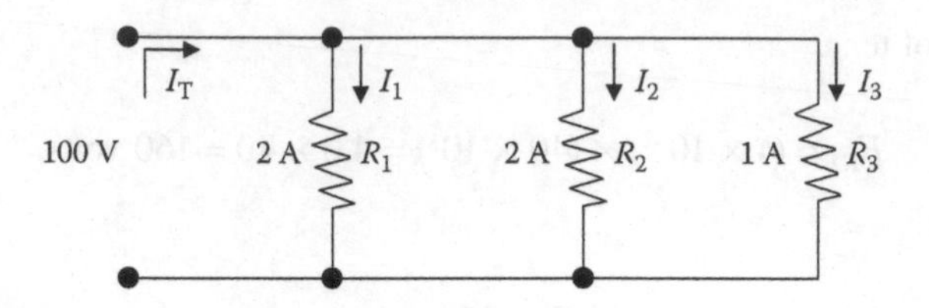

FIGURE 10.22 Illustration for Example 10.18.

By Ohm's law, each branch current equals the applied voltage divided by the resistance between the two points where the voltage is applied. Hence, for each branch we have the following equations:

$$\begin{aligned}
&\text{Branch 1:} \quad I_1 = \frac{E_1}{R_1} = \frac{V}{R_1} \\
&\text{Branch 2:} \quad I_2 = \frac{E_2}{R_2} = \frac{V}{R_2} \\
&\text{Branch 3:} \quad I_3 = \frac{E_3}{R_3} = \frac{V}{R_3}
\end{aligned} \tag{10.14}$$

With the same applied voltage, any branch that has less resistance allows more current through it than a branch with higher resistance.

■ Example 10.18

Problem: Two resistors, each drawing 2 amperes, and a third resistor that draws 1 ampere are connected in parallel across a 100-volt line (see Figure 10.22). What is the total current?

Solution: The formula for total current is

$$I_T = I_1 + I_2 + I_3$$

Thus,

$$I_T = 2 \text{ amperes} + 2 \text{ amperes} + 1 \text{ amperes} = 5 \text{ amperes}$$

The total current, then, is 5 amperes.

■ Example 10.19

Problem: Two branches, R_1 and R_2, across a 100-volt power line draw a total line current of 20 amperes (Figure 10.23). Branch R_1 takes 10 amperes. What is the current (I_2) in branch R_2?

Solution: Beginning with Equation 10.13, transpose to find I_2 and then substitute given values:

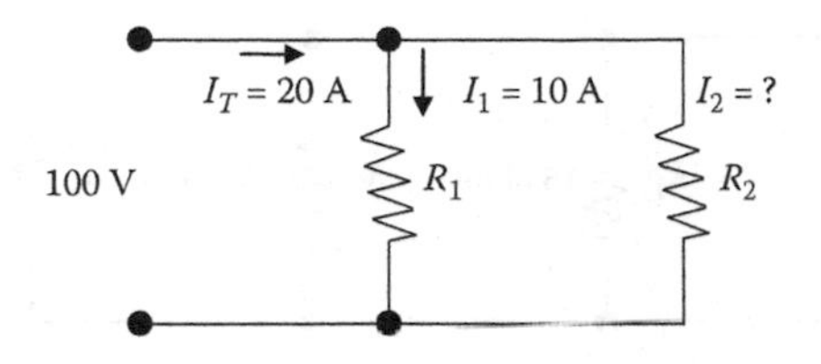

FIGURE 10.23 Illustration for Example 10.19.

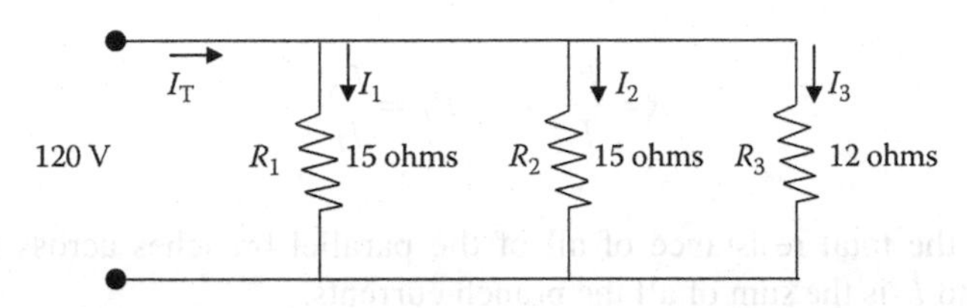

FIGURE 10.24 Illustration for Example 10.20.

$$I_T = I_1 + I_2$$

$$I_2 = I_T - I_1 = 20 \text{ amperes} - 10 \text{ amperes} = 10 \text{ amperes}$$

The current in branch R_2, then, is 10 amperes.

■ Example 10.20

Problem: A parallel circuit consists of two 15-ohm and one 12-ohm resistors across a 120-volt line (see Figure 10.24). What current will flow in each branch of the circuit and what is the total current drawn by all the resistors?

Solution: There is a 120-volt potential across each resistor. Using Equation 10.13, apply Ohm's law to each resistor:

$$I_1 = \frac{V}{R_1} = \frac{120}{15} = 8 \text{ amperes}$$

$$I_2 = \frac{V}{R_2} = \frac{120}{15} = 8 \text{ amperes}$$

$$I_3 = \frac{V}{R_3} = \frac{120}{12} = 10 \text{ amperes}$$

Parallel Circuit Resistance

Unlike series circuits, where total resistance (R_T) is the sum of the individual resistances, in a parallel circuit the total resistance is *not* the sum of the individual resistances. In a parallel circuit, we can use Ohm's law to find total resistance:

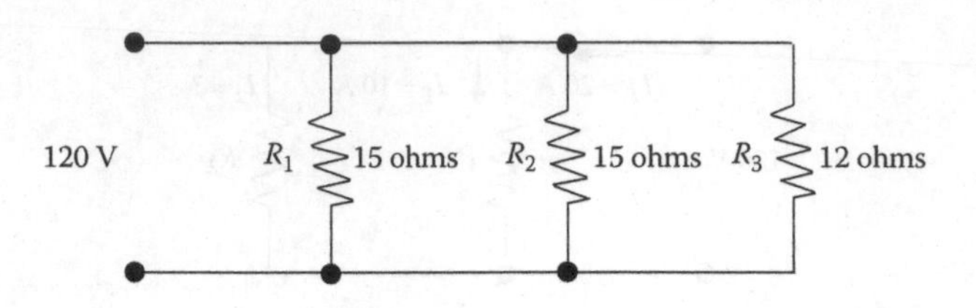

FIGURE 10.25 Illustration for Example 10.21.

$$R = \frac{E}{I} \quad \text{or} \quad R_T = \frac{E_S}{I_T}$$

where R_T is the total resistance of all of the parallel branches across the voltage source E_S, and I_T is the sum of all the branch currents.

■ Example 10.21

Problem: Given that $E_S = 120$ volts and $I_T = 26$ amperes, what is the total resistance of the circuit shown in Figure 10.25?

Solution: In Figure 10.25, the line voltage is 120 volts and the total line current is 26 amperes; therefore,

$$R_T = \frac{E_S}{I_T} = \frac{120}{26} = 4.62 \text{ ohms}$$

Other methods are used to determine the equivalent resistance of parallel circuits. The most appropriate method for a particular circuit depends on the number and value of the resistors; for example, consider the parallel circuit shown in Figure 10.26. For this circuit, the following simple equation is used:

$$R_{eq} = \frac{R}{N} \tag{10.15}$$

where

R_{eq} = Equivalent parallel resistance.
R = Ohmic value of one resistor.
N = Number of resistors.

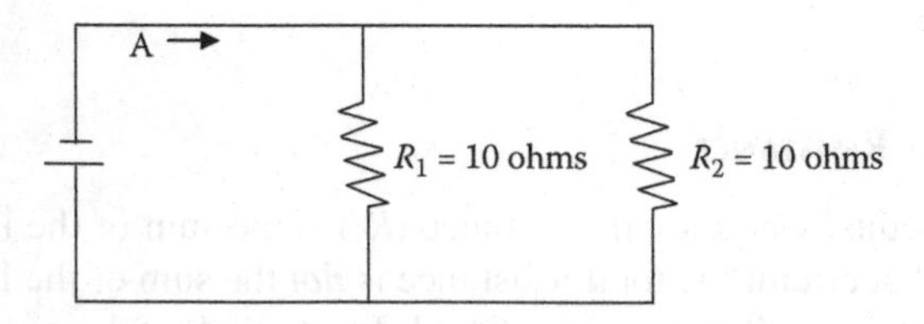

FIGURE 10.26 Two equal resistors connected in parallel.

Thus,

$$R_{eq} = \frac{R}{N} = \frac{10 \text{ ohms}}{2} = 5 \text{ ohms}$$

Note: When two equal value resistors are connected in parallel, they present a total resistance equivalent to a single resistor of one-half the value of either of the original resistors.

■ Example 10.22

Problem: Five 50-ohm resistors are connected in parallel. What is the equivalent circuit resistance?

Solution:

$$R_{eq} = \frac{R}{N} = \frac{50 \text{ ohms}}{5} = 10 \text{ ohms}$$

What about parallel circuits containing resistance of unequal value? How is equivalent resistance determined? Example 10.23 demonstrates how this is accomplished.

■ Example 10.23

Problem: Refer to Figure 10.27.

Solution:

Given:

$R_1 = 3$ ohms
$R_2 = 6$ ohms
$E_a = 30$ volts

We know that

$I_1 = 10$ amperes
$I_2 = 5$ amperes
$I_T = 15$ amperes

and can now determine R_{eq}:

$$R_{eq} = \frac{E_a}{I_T} = \frac{30}{15} = 2 \text{ ohms}$$

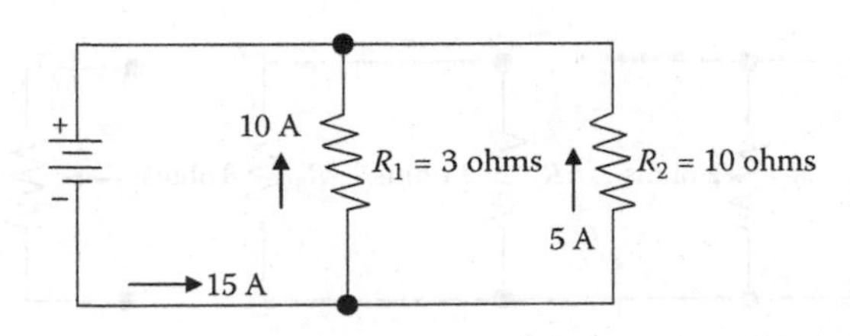

FIGURE 10.27 Illustration for Example 10.23.

Reciprocal Method

When circuits are encountered in which resistors of unequal value are connected in parallel, the equivalent resistance may be computed by using the *reciprocal method.*

> ***Note:*** A *reciprocal* is an inverted fraction; the reciprocal of the fraction 3/4, for example, is 4/3. We consider a whole number to be a fraction with 1 as the denominator, so the reciprocal of a whole number is that number divided into 1; for example, the reciprocal of R_T is $1/R_T$. The equivalent resistance in parallel is given by the following formula:

$$\frac{1}{R_T} = \frac{1}{R_1} + \frac{1}{R_2} + \frac{1}{R_3} + \ldots + \frac{1}{R_n} \tag{10.16}$$

where R_T is the total resistance in parallel, and R_1, R_2, R_3, and R_n are the branch resistances.

■ Example 10.24

Problem: Find the total resistance of a 2-ohm, a 4-ohm, and an 8-ohm resistor in parallel (Figure 10.28).

Solution: Write the formula for the three resistors in parallel:

$$\frac{1}{R_T} = \frac{1}{R_1} + \frac{1}{R_2} + \frac{1}{R_3}$$

Substitute the resistance values:

$$\frac{1}{R_T} = \frac{1}{2} + \frac{1}{4} + \frac{1}{8}$$

Add the fractions:

$$\frac{1}{R_T} = \frac{4}{8} + \frac{2}{8} + \frac{1}{8} = \frac{7}{8}$$

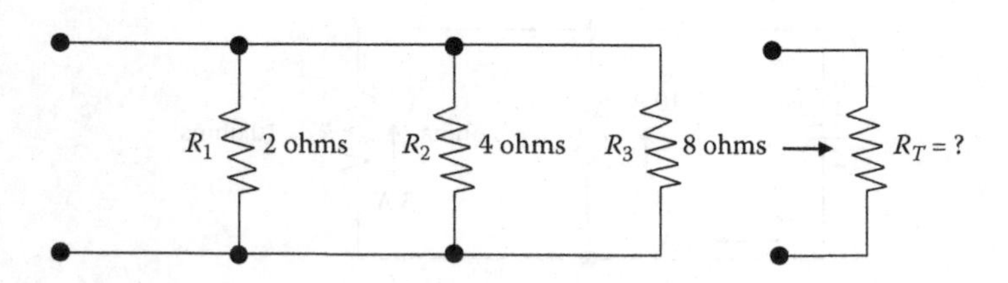

FIGURE 10.28 Illustration for Example 10.24.

Invert both sides of the equation to solve for R_T:

$$R_T = \frac{8}{7} = 1.14 \text{ ohms}$$

Note: When resistances are connected in parallel, the total resistance is always less than the smallest resistance of any single branch.

Product over the Sum Method

When any two unequal resistors are in parallel, it is often easier to calculate the total resistance by multiplying the two resistances and then dividing the product by the sum of the resistances:

$$R_T = \frac{R_1 \times R_2}{R_1 + R_2} \quad (10.17)$$

where R_T is the total resistance in parallel, and R_1 and R_2 are the two resistors in parallel.

■ Example 10.25

Problem: What is the equivalent resistance of a 20-ohm and a 30-ohm resistor connected in parallel?

Solution:

Given:

$R_1 = 20$ ohms
$R_2 = 30$ ohms

$$R_T = \frac{R_1 \times R_2}{R_1 + R_2} = \frac{20 \times 30}{20 + 30} = 12 \text{ ohms}$$

Power in Parallel Circuits

As in the series circuit, the total power consumed in a parallel circuit is equal to the sum of the power consumed in the individual resistors.

Note: Because power dissipation in resistors consists of a heat loss, power dissipations are additive regardless of how the resistors are connected in the circuit

$$P_T = P_1 + P_2 + P_3 + \ldots + P_n \quad (10.18)$$

where P_T is the total power, and P_1, P_2, P_3, … P_n are the branch powers.

Total power can also be calculated by the following equation:

$$P_T = E \times I_T \quad (10.19)$$

where P_T is the total power, E is the voltage source across all parallel branches, and I_T is the total current. The power dissipated in each branch is equal to $E \times I$ and equal to V^2/R.

> ***Note:*** In both parallel and series arrangements, the sum of the individual values of power dissipated in the circuit equals the total power generated by the source. The circuit arrangements cannot change the fact that all of the power in the circuit comes from the source.

11 Circumference, Area, and Volume

$$\text{Area} = L \times W \quad \text{Area of circle} = 0.785 \times D^2 \quad \text{Volume} = L \times W \times D$$

Water/wastewater treatment plants consist of a series of tanks and channels. Proper operational control requires the operator to perform several process control calculations involving parameters such as the circumference, perimeter, area, or volume of a tank or channel. Many process calculations require computation of surface areas. To aid in performing these calculations, the following definitions are provided:

Area—The area of an object, measured in square units.
Base—The bottom leg of a triangle, measured in linear units.
Circumference—The distance around an object, measured in linear units. When determined for other than circles, it may be called the *perimeter* of the figure, object, or landscape.
Cubic units—Measurements used to express volume (e.g., cubic feet, cubic meters).
Depth—The vertical distance from the bottom the tank to the top. It is normally measured in terms of liquid depth and given in terms of sidewall depth (SWD), measured in linear units.
Diameter—The distance, measured in linear units, from one edge of a circle to the opposite edge passing through the center.
Height—The vertical distance, measured in linear units, from one end of an object to the other.
Length—Distance, measured in linear units, from one end of an object to the other.
Linear units—Measurements used to express distance (e.g., feet, inches, meters, yards).
Pi (π)—A number in the calculations involving circles, spheres, or cones ($\pi = 3.14$).
Radius—Distance, measured in linear units, from the center of a circle to the edge.
Sphere—A container shaped like a ball.
Square units—Measurements used to express area (e.g., square feet, square meters, acres).
Volume—The capacity of a unit (how much it will hold), measured in cubic units (e.g., cubic feet, cubic meters) or in liquid volume units (e.g., gallons, liters, million gallons).
Width—Distance from one side of a tank to the other, measured in linear units.

In addition to understanding the above terminology used in this chapter for measurements of circumference, area, and volume, it is also important to have a strategy for solving these types of problems as explained in the following:

- Always read the problem, disregard the numbers. Ask yourself: What type of problem is it? What am I asked to find?
- If a diagram is provided, refer to it. If there isn't one, draw a rough one to refer to (see Figure 11.1 for an example).
- What exactly do I need to know to solve the problem; that is, how is the information presented in the statement of the problem?
- Write out everything that is known about the problem in one column and place the unknown in another column.
- Identify the correct formula and plug in the numbers and solve.
- Work out the problem.
- Check the answer; does it make sense? Make sure the measurements agree. If the diameter of a pipe is given in inches, then change the inches to feet; if the flow is in MGD and you need a value in feet or feet/sec, then change MGD to ft^3/sec.

Let's look at Example 11.1, where we put the steps listed above to work to solve an area problem.

■ Example 11.1

Problem: A water storage basin is 30 ft in length and 45 ft in width. What is the area of the basin in square feet? (Refer to Figure 11.1.)

Solution:

Known
 Length = 30 ft
 Width = 45 ft

Unknown
 Area = ?

$$\text{Area} = \text{Length} \times \text{Width} = 30 \text{ ft} \times 45 \text{ ft} = 1350 \text{ ft}^2$$

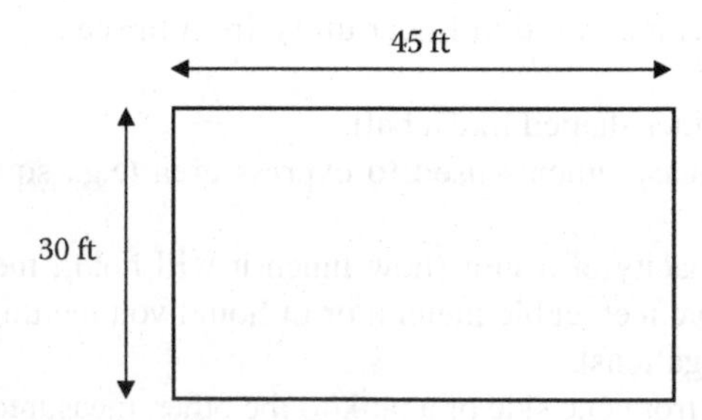

FIGURE 11.1 Illustration for Example 11.1.

PERIMETER AND CIRCUMFERENCE

On occasion, it may be necessary to determine the distance around grounds or landscapes. To measure the distance around property, buildings, and basin-like structures, it is necessary to determine either perimeter or circumference. The *perimeter* is the distance around an object; it is the border or outer boundary. *Circumference* is the distance around a circle or circular object, such as a clarifier. Distance is a linear measurement that defines the distance (or length) along a line. Standard units of measurement such as inches, feet, yards, and miles and metric units such as centimeters, meters, and kilometers are used.

PERIMETER

The perimeter (P) of a rectangle (a four-sided figure with four right angles) is obtained by adding the lengths (L_i) of the four sides (see Figure 11.2):

$$\text{Perimeter} = L_1 + L_2 + L_3 + L_4 \tag{11.1}$$

■ **EXAMPLE 11.2**

Problem: Find the perimeter of the rectangle shown in Figure 11.3.

Solution:

$$\text{Perimeter} = 35 \text{ ft} + 8 \text{ ft} + 35 \text{ ft} + 8 \text{ ft} = 86 \text{ ft}$$

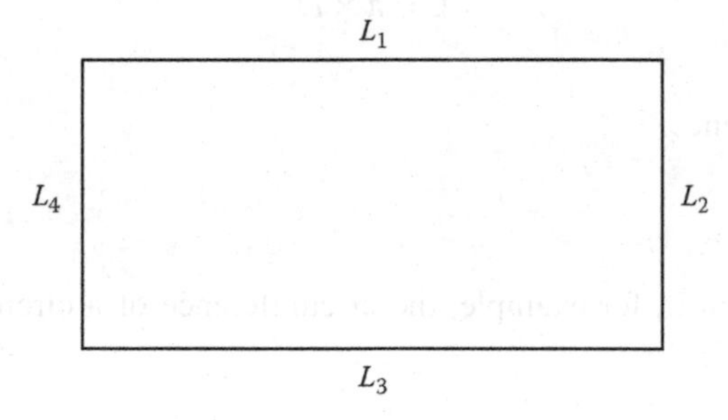

FIGURE 11.2 Perimeter.

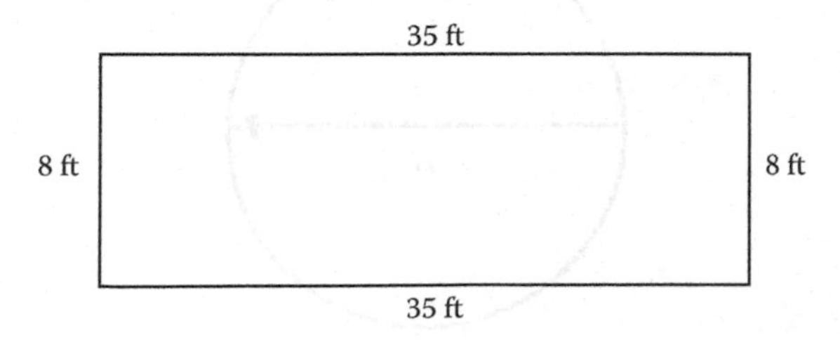

FIGURE 11.3 Perimeter of a rectangle for Example 11.2.

■ Example 11.3

Problem: What is the perimeter of a rectangular field if its length is 100 ft and its width is 50 ft?

Solution:

$$\text{Perimeter} = (2 \times \text{Length}) + (2 \times \text{Width})$$
$$= (2 \times 100\text{ ft}) + (2 \times 50\text{ ft}) = 200\text{ ft} + 100\text{ ft} = 300\text{ ft}$$

■ Example 11.4

Problem: What is the perimeter of a square with 8-in. sides?

Solution:

$$\text{Perimeter} = (2 \times \text{Length}) + (2 \times \text{Width})$$
$$= (2 \times 8\text{ in.}) + (2 \times 8\text{ in.}) = 16\text{ in.} + 16\text{ in.} = 32\text{ in.}$$

Circumference

The circumference is the length of the outer border of a circle. The circumference is found by multiplying pi (π) times the *diameter* (D) (a straight line passing through the center of a circle, or the distance across the circle; see Figure 11.4):

$$C = \pi \times D \tag{11.2}$$

where
C = Circumference.
π = pi = 3.14.
D = Diameter.

Use this calculation if, for example, the circumference of a circular tank must be determined.

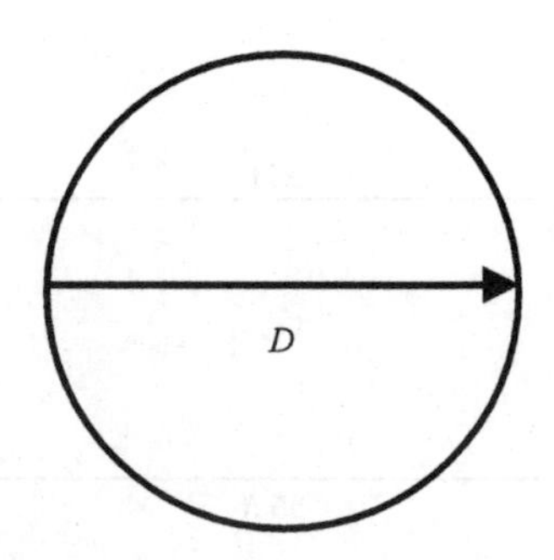

FIGURE 11.4 Circumference.

■ EXAMPLE 11.5

Problem: Find the circumference of a circle that has a diameter of 25 ft ($\pi = 3.14$).

Solution:

$$C = \pi \times D = 3.14 \times 25 \text{ ft} = 78.5 \text{ ft}$$

■ EXAMPLE 11.6

Problem: A circular chemical holding tank has a diameter of 18 meters (m). What is the circumference of this tank?

Solution:

$$C = \pi \times D = 3.14 \times 18 \text{ m} = 56.52 \text{ m}$$

■ EXAMPLE 11.7

Problem: An influent pipe inlet opening has a diameter of 6 ft. What is the circumference of the inlet opening?

Solution:

$$C = \pi \times D = 3.14 \times 6 \text{ ft} = 18.84 \text{ ft}$$

AREA

For area measurements in water/wastewater operations, three basic shapes are particularly important—namely, circles, rectangles, and triangles. Area is the amount of surface an object contains or the amount of material it takes to cover the surface. The area on top of a chemical tank is called the *surface area.* The area of the end of a ventilation duct is called the *cross-sectional area* (the area at right angles to the length of ducting). Area is usually express in square units, such as square inches (in.^{2}) or square feet (ft^2). Land may also be expressed in terms of square miles (sections) or acres (43,560 ft^2) or, in the metric system, as hectares (2.47 acres, 10,000 m^2, or 107,600 ft^2).

AREA OF A RECTANGLE

A rectangle is a two-dimensional box. The area of a rectangle is found by multiplying the length (L) times the width (W) (see Figure 11.5).

$$\text{Area} = L \times W \tag{11.3}$$

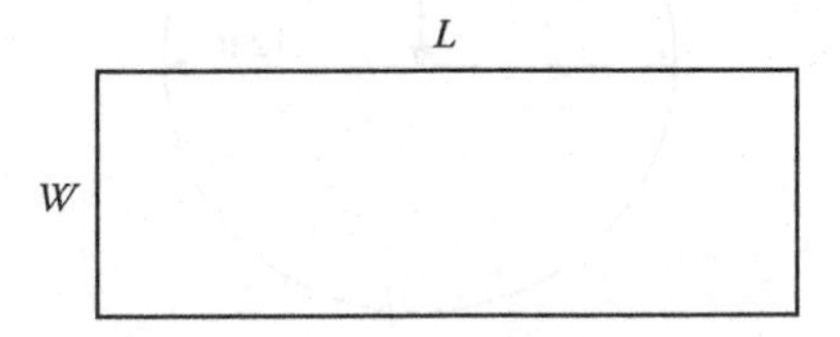

FIGURE 11.5 Area of a rectangle.

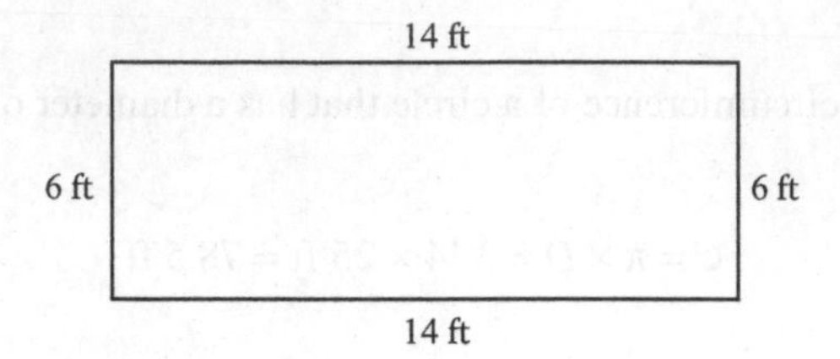

FIGURE 11.6 Illustration for Example 11.8.

■ EXAMPLE 11.8

Problem: Find the area of the rectangle shown in Figure 11.6.

Solution:

$$\text{Area} = L \times W = 14\text{ ft} \times 6\text{ ft} = 84\text{ ft}^2$$

AREA OF A CIRCLE

In order to calculate volumes of circular tanks and velocities in pipes, the area of the circle must first be determined. There are two basic formulas used to calculate the area of a circle.

$$\text{Area of circle} = \pi \times r^2 \tag{11.4}$$

$$\text{Area of circle} = 0.785 \times D^2 \tag{11.5}$$

where
π = pi = 3.14.
r = Radius of circle = one-half of the diameter.
D = Diameter.

The radius (r) is any straight line that radiates form the center of the circle to some point on the circumference. By definition, all radii of the same circle are equal. The area of a circle is determined by multiplying π times the radius squared. In Figure 11.7, we have a circle with a radius of 6 inches.

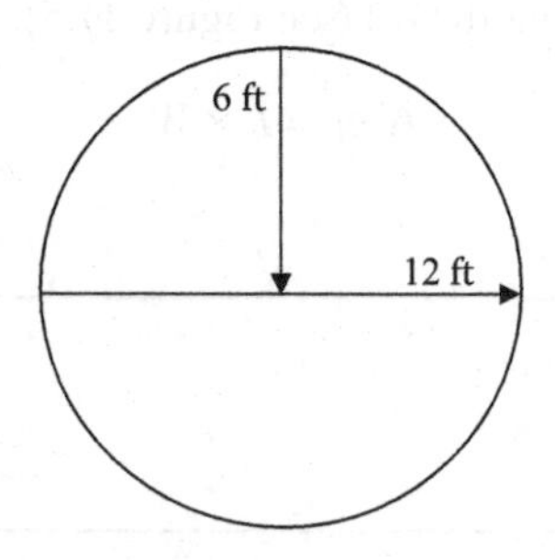

FIGURE 11.7 Illustration of area of a circle.

■ EXAMPLE 11.9

Problem: What is the area of the circle shown in Figure 11.7?

Solution:

$$\text{Area of circle} = \pi \times r^2 = \pi \times 6^2 = 3.14 \times 36 = 113 \text{ ft}^2$$

AREA OF A CIRCULAR OR CYLINDRICAL TANK

If we were assigned to paint a water storage tank, we must know the surface area of the walls of the tank so we can determine how much paint is required for the job. To compute the surface area of the tank, we should first visualize the cylindrical walls as being a rectangle wrapped around a circular base. The area of a rectangle is found by multiplying the length by the width; in this case, the width of the rectangle is the height of the wall, and the length of the rectangle is the distance around the circle—that is, the circumference. Thus, the area of the side walls of the circular tank is found by multiplying the circumference of the base ($C = \pi \times D$) times the height of the wall (H):

$$\text{Area} = \pi \times D \times H \tag{11.6}$$

■ EXAMPLE 11.10

Problem: A cylindrical tank is to be painted. The top surface area of the tank is 314 ft². The diameter is 20 ft and the height is 25 ft. How much paint is needed to cover the tank? (See Figure 11.8.)

Solution:

$$\text{Area} = \pi \times 20 \text{ ft} \times 25 \text{ ft} = 3.14 \times 20 \text{ ft} \times 25 \text{ ft} = 1570.8 \text{ ft}^2$$

To determine the amount of paint needed, remember to add the surface area of the top of the tank, which is 314 ft². Thus, there must be sufficient paint to cover 1570.8 ft² + 314 ft² = 1884.8 or 1885 ft². If the tank floor should be painted, add another 314 ft².

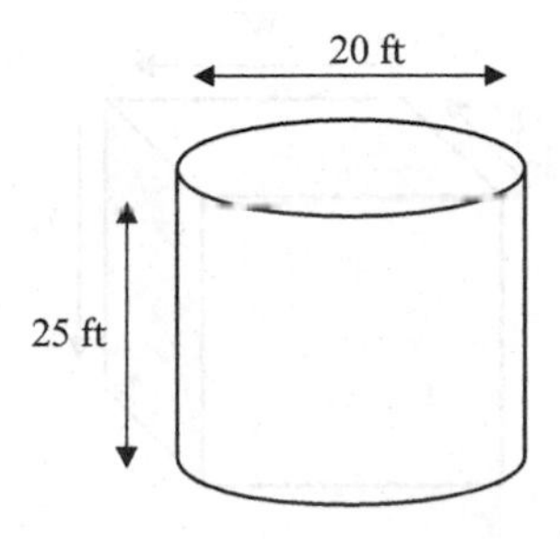

FIGURE 11.8 Illustration of cylindrical tank for Example 11.10.

■ **Example 11.11**

Problem: A sedimentation basin is 70 ft in diameter. What is the surface area of the tank?

Solution:

$$\text{Area} = \pi \times r^2 = 3.14 \times (35 \text{ ft} \times 35 \text{ ft}) = 3850 \text{ ft}^2$$

$$\text{Area} = 0.785 \times D^2 = 0.785 \times (70 \text{ ft} \times 70 \text{ ft}) = 3850 \text{ ft}^2$$

■ **Example 11.12**

Problem: A pipeline has a diameter of 14 in. What is the area of the pipe?

Solution:

$$\text{Area} = \pi \times r^2 = 3.14 \times (7 \text{ in.} \times 7 \text{ in.}) = 154 \text{ in.}^2$$

$$\text{Area} = 0.785 \times D^2 = 0.785 \times (14 \text{ in.} \times 14 \text{ in.}) = 154 \text{ in.}^2$$

VOLUME

Volume is the amount of space occupied by or contained in an object (see Figure 11.9). It is expressed in cubic units, such as cubic inches (in.^{3}), cubic feet (ft^3), or acre-feet (1 acre-foot = 43,560 ft^3).

Volume of a Rectangular Basin

The volume (V) of a rectangular object is obtained by multiplying the length by the width by the depth or height:

$$V = L \times W \times H \tag{11.7}$$

where:

L = Length.
W = Width.
H = Height (or depth).

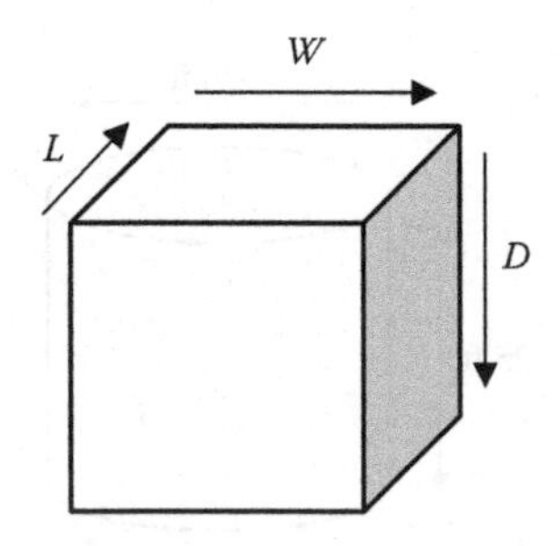

FIGURE 11.9 Volume.

TABLE 11.1
Volume Formulas

Sphere volume	=	$(\pi/6) \times (\text{Diameter})^3$
Cone volume	=	1/3 × Volume of a cylinder
Rectangular tank volume	=	Area of rectangle × H (or depth)
	=	$L \times W \times H$ (or depth)
Cylinder volume	=	Area of cylinder × H (or depth)
	=	$\pi \times r^2 \times H$ (or depth)

■ Example 11.13

Problem: A unit rectangular process basin has a length of 15 ft, width of 7 ft, and depth of 9 ft. What is the volume of the basin?

Solution:

$$V = L \times W \times D = 15 \text{ ft} \times 7 \text{ ft} \times 9 \text{ ft} = 945 \text{ ft}^3$$

■ Example 11.14

Problem: A sedimentation tank is 60 ft long, 40 ft wide, and 12 ft deep. What is the volume of the tank in cubic feet?

Solution:

$$60 \text{ ft} \times 40 \text{ ft} \times 12 \text{ ft} = 28{,}800 \text{ ft}^3$$

For wastewater operators, representative surface areas are most often rectangles, triangles, circles, or a combination of these. Practical volume formulas used in water/wastewater calculations are given in Table 11.1.

Volume of Round Pipe and Round Surface Areas

To determine the volume of round pipe and round surface areas, the following examples are helpful.

■ Example 11.15

Problem: Find the volume of a 3-in. round pipe that is 300 ft long.

Solution:

1. Change the diameter of the duct from inches to feet by dividing 3 in. by 12:

$$3 \text{ in.} \div 12 \text{ in./ft} = 0.25 \text{ ft}$$

2. Find the radius (r) by dividing the diameter by 2:

$$0.25 \text{ ft} \div 2 = 0.125$$

3. Find the volume (V):

$$V = L \times \pi \times r^2$$

$$V = 300 \text{ ft} \times 3.14 \times 0.0156 = 14.7 \text{ ft}^2$$

■ Example 11.16

Problem: Find the volume of a smokestack that is 24 in. in diameter (entire length) and 96 in. tall.

Solution: First find the radius of the stack. The radius is one half the diameter, so 24 in. ÷ 2 = 12 in. Now find the volume:

$$V = H \times \pi \times r^2 = 96 \text{ in.} \times 3.14 \times (12 \text{ in.})^2 = 96 \text{ in.} \times 3.14 \times 144 \text{ in.}^2 = 43{,}407 \text{ ft}^3$$

■ Example 11.17

Problem: A sedimentation basin is 80 ft in diameter and 12 ft deep. What is the volume of the tank?

Solution:

$$V = H \times \pi \times r^2 = 12 \text{ ft} \times 3.14 \times (40 \text{ ft} \times 40 \text{ ft}) = 60{,}403 \text{ ft}^3$$

Volume of a Cone and Sphere

The following equations and examples show how to determine the volume of a cone, a sphere, and a tank.

Volume of a Cone

$$\text{Volume of a cone} = (\pi/12) \times (\text{Diameter})^2 \times \text{Height} \quad (11.8)$$

$$(\pi/12) = (3.14/12) = 0.262$$

Note: The diameter used in the formula is the diameter of the base of the cone.

■ Example 11.18

Problem: The bottom section of a circular settling tank has the shape of a cone. How many cubic feet of water are contained in this section of the tank if the tank has a diameter of 120 ft and the cone portion of the unit has a depth of 6 ft?

Solution:

$$V = 0.262 \times (120 \text{ ft})^2 \times 6 \text{ ft} = 22{,}637 \text{ ft}^3$$

Volume of a Sphere

$$\text{Volume of a sphere} = (\pi/6) \times (\text{Diameter})^3 \quad (11.9)$$

$$(\pi/6) = (3.14/6) = 0.524$$

■ Example 11.19

Problem: What is the volume in cubic feet of a gas storage container that is spherical and has a diameter of 60 ft?

Solution:

$$V = 0.524 \times (60 \text{ ft})^3 = 113{,}184 \text{ ft}^3$$

Volume of a Circular or Cylindrical Tank

Circular process and various water and chemical storage tanks are commonly found in water/wastewater treatment. A circular tank consists of a circular floor surface with a cylinder rising above it (see Figure 11.10). The volume of a circular tank is calculated by multiplying the surface area times the height of the tank walls.

■ Example 11.20

Problem: If a tank is 20 ft in diameter and 25 ft deep, how many gallons of water will it hold?

Hint: In this type of problem, calculate the surface area first, multiply by the height, and then convert to gallons.

Solution:

$$r = D \div 2 = 20 \text{ ft} \div 2 = 10 \text{ ft}$$
$$A = \pi \times r^2 = 3.14 \times (10 \text{ ft})^2 = 314 \text{ ft}^2$$
$$V = A \times H = 314 \text{ ft}^2 \times 25 \text{ ft} = 7850 \text{ ft}^3$$
$$7850 \text{ ft}^3 \times 7.48 \text{ gal/ft}^3 = 58{,}718 \text{ gal}$$

Volume in Gallons

It is often necessary to calculate a volume of a tank or pipe in gallons rather than cubic feet. In most cases the volume must be calculated in cubic feet and then converted into gallons. This is determined by multiplying cubic feet by 7.48.

$$\text{Volume in gallons} = \text{Cubic feet} \times 7.48 \text{ gal/ft}^3$$

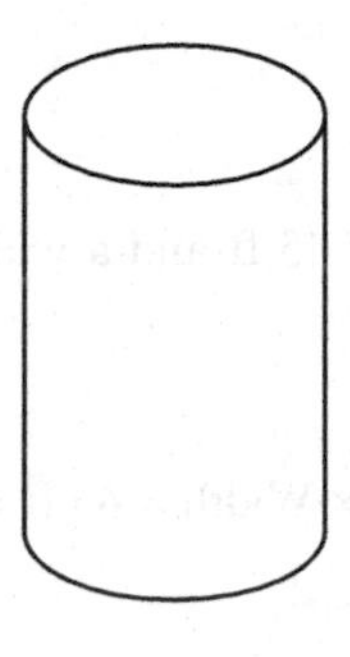

FIGURE 11.10 Cylindrical tank.

■ Example 11.21

Problem: A sedimentation basin is 60 ft long by 40 ft wide and 12 ft deep. What is the volume of the tank in gallons?

Solution:

$$V = A \times H = 60\ \text{ft} \times 40\ \text{ft} \times 12\ \text{ft} = 28{,}800\ \text{ft}^3$$

$$28{,}800\ \text{ft}^3 \times 7.48\ \text{gal/ft}^3 = 215{,}425\ \text{gal}$$

■ Example 11.22

Problem: A circular tank has a diameter of 40 ft and is 12 ft deep. How many gallons will it hold?

Solution:

$$V = A \times H = (40\ \text{ft})^2 \times 0.785 \times 12\ \text{ft} = 15{,}072\ \text{ft}^3$$

$$15{,}072\ \text{ft}^3 \times 7.48\ \text{gal/ft}^3 = 112{,}739\ \text{gal}$$

Volume of Pipes

The number of gallons contained in a 1-ft section of pipe can be determined by squaring the diameter (in inches) and then multiplying by 0.048. To determine the number of gallons in a particular length of pipe, multiply the gallons per foot by the number of feet of pipe.

$$\text{Volume in gallons} = D^2\ (\text{in.}) \times 0.0408 \times \text{Length (ft)} \tag{11.10}$$

■ Example 11.23

Problem: A 12-in. line is 1200 ft long. How many gallons does the pipe hold?
Solution:

$$V = (12\ \text{in.})^2 \times 0.0408 \times 1200\ \text{ft} = 7050\ \text{gal}$$

AREA EXAMPLES

■ Example 11.24

Problem: A basin has a length of 45 ft and a width of 14 ft. Calculate the area in square feet.

Solution:

$$\text{Area} = \text{Length} \times \text{Width} = 45\ \text{ft} \times 14\ \text{ft} = 630\ \text{ft}^2$$

■ Example 11.25

Problem: Calculate the surface area of a basin that is 90 ft long, 20 ft wide, and 12 ft deep.

Solution:

$$\text{Area} = \text{Length} \times \text{Width} = 90 \text{ ft} \times 20 \text{ ft} = 1800 \text{ ft}^2$$

■ Example 11.26

Problem: Calculate the cross-sectional area in square feet for a 3-ft-diameter main that has just been laid.

Solution:

$$\text{Area} = 0.785 \times D^2 = 0.785 \times (3 \text{ ft})^2 = 7.1 \text{ ft}^2$$

■ Example 11.27

Problem: Calculate the cross-sectional area in square feet for a 36-in.-diameter main that has just been laid.

Solution:

$$36 \text{ in.} = 3 \text{ ft}$$

$$A = 0.785 \times D^2 = 0.785 \times (3 \text{ ft})^2 = 7.1 \text{ ft}^2$$

■ Example 11.28

Problem: Calculate the area in square feet for a 48-in. interceptor line that has just been laid.

Solution:

$$48 \text{ in.} = 4 \text{ ft}$$

$$A = 0.785 \times D^2 = 0.785 \times (4 \text{ ft})^2 = 12.6 \text{ ft}^2$$

■ Example 11.29

Problem: Calculate the cross-sectional area in square feet for a 3-in. pipe.

Solution:

$$3 \text{ in.} = 0.25 \text{ ft}$$

$$A = 0.785 \times D^2 = 0.785 \times (0.25 \text{ ft})^2 = 0.049 \text{ ft}^2$$

VOLUME EXAMPLES

■ Example 11.30

Problem: Calculate the volume in cubic feet of a tank that measures 12 ft by 12 ft by 12 ft.

Solution:

$$\text{Volume} = L \times W \times D = 12 \text{ ft} \times 12 \text{ ft} \times 12 \text{ ft} = 1728 \text{ ft}^3$$

■ Example 11.31

Problem: Calculate the volume in gallons for a basin that measures 24 ft by 12 ft by 4 ft.

Solution:

$$\text{Volume} = L \times W \times D = 24 \text{ ft} \times 12 \text{ ft} \times 4 \text{ ft} = 1152 \text{ ft}^3$$

$$1152 \text{ ft}^3 \times 7.48 \text{ ft}^3\text{/gal} = 8616.9 \text{ gal}$$

■ Example 11.32

Problem: Calculate the volume in gallons of water in a tank that is 266 ft long, 66 ft wide, and 12 ft deep if the tank contains 3 ft of water.

Solution:

$$\text{Volume} = L \times W \times D = 266 \text{ ft} \times 66 \text{ ft} \times 3 \text{ ft} = 52{,}668 \text{ ft}^3$$

$$52{,}668 \text{ ft}^3 \times 7.48 \text{ ft}^3\text{/gal} = 393{,}956.64 \text{ gal}$$

■ Example 11.33

Problem: Calculate the volume of water in a tank in gallons that measures 14 ft long, 8 ft wide, and 4 ft deep and contains 6 in. of water

Solution:

$$6 \text{ in.} = 0.5 \text{ ft}$$

$$\text{Volume} = L \times W \times D = 14 \text{ ft} \times 4 \text{ ft} \times 0.5 \text{ ft} = 28 \text{ ft}^3$$

$$28 \text{ ft}^3 \times 7.48 \text{ ft}^3\text{/gal} = 209.4 \text{ gal}$$

■ Example 11.34

Problem: Calculate the maximum volume of water in gallons for a kids' swimming pool that measures 8 ft across and can hold 18 in. of water.

Solution:

$$18 \text{ in.} = 1.5 \text{ ft}$$

$$\text{Volume} = 0.785 \times D^2 \times \text{Height} = 0.785 \times (8 \text{ ft})^2 \times 1.5 \text{ ft} = 75.36 \text{ ft}^3$$

$$75.36 \text{ ft}^3 \times 7.48 \text{ gal/ft}^3 = 564.45 \text{ gal}$$

■ **EXAMPLE 11.35**

Problem: How many gallons of water can a barrel hold if it measures 4 ft in diameter and can hold water to a depth of 4 ft?

Solution:

$$\text{Volume} = 0.785 \times D^2 \times \text{Height} = 0.785 \times (4\ \text{ft})^2 \times 4\ \text{ft} = 50.24\ \text{ft}^3$$

$$50.24\ \text{ft}^3 \times 7.48\ \text{gal/ft}^3 = 375.8\ \text{gal}$$

■ **EXAMPLE 11.36**

Problem: A water main needs to be disinfected. The main is 42 inches in diameter and has a length of 0.33 miles. How many gallons of water will it hold?

Solution:

$$42\ \text{in.} = 3.5\ \text{ft}$$

$$0.33\ \text{mile} \times 5280\ \text{ft/mile} = 1742.4\ \text{ft}$$

$$\text{Volume} = 0.785 \times D^2 \times \text{Height} = 0.785 \times (3.5\ \text{ft})^2 \times 1742.4\ \text{ft} = 16{,}755.35\ \text{ft}^3$$

$$16{,}755.35\ \text{ft}^3 \times 7.48\ \text{gal/ft}^3 = 125{,}330.04\ \text{gal}$$

■ **EXAMPLE 11.37**

Problem: A 5-million-gallon storage tank is 5% full. How many gallons does it contain?

Solution:

$$5{,}000{,}000\ \text{gal} \times 0.05 = 250{,}000\ \text{gal}$$

■ **EXAMPLE 11.38**

Problem: What is 11% of a 1.5-million-gallon tank?

Solution:

$$1.5\ \text{MG} \times 0.11 = 0.165\ \text{MG}$$

12 Force, Pressure, Head, and Velocity Calculations

When there is a change in pressure at any point in a fluid, the change in pressure is transmitted equally and unchanged in all directions through the fluid.

—Pascal's law

Before we study calculations involving the relationship between force, pressure, and head, we must first define these terms:

- *Force*—The push exerted by water on any confining surface. Force can be expressed in pounds, tons, grams, or kilograms.
- *Pressure*–The force per unit area. The most common way of expressing pressure is in pounds per square inch (psi).
- *Head*—The vertical distance or height of water above a reference point. Head is usually expressed in feet. In the case of water, head and pressure are related.

Note: Water pressure is measured in terms of pounds per square inch (psi) and feet of head (height of a water column) in feet. A column of water 2.31 ft high creates a pressure of 1 psi.

FORCE AND PRESSURE

Figure 12.1 illustrates these terms. A cubical container measuring 1 foot on each side can hold 1 cubic foot of water. A basic fact of science states that 1 cubic foot of water weighs 62.4 pounds. The force acting on the bottom of the container would be 62.4 pounds, and the pressure acting on the bottom of the container would be 62.4 pounds per square foot (lb/ft^2, psf). The area of the bottom in square inches is

$$1\text{ ft}^2 = 12\text{ in.} \times 12\text{ in.} = 144\text{ in.}^2$$

Therefore, the pressure in pounds per square inch (psi) is

$$\frac{62.4\text{ lb/ft}^2}{1\text{ ft}^2} = \frac{62.4\text{ lb/ft}^2}{144\text{ in.}^2/\text{ft}^2} = 0.433\text{ lb/in.}^2\text{ (psi)}$$

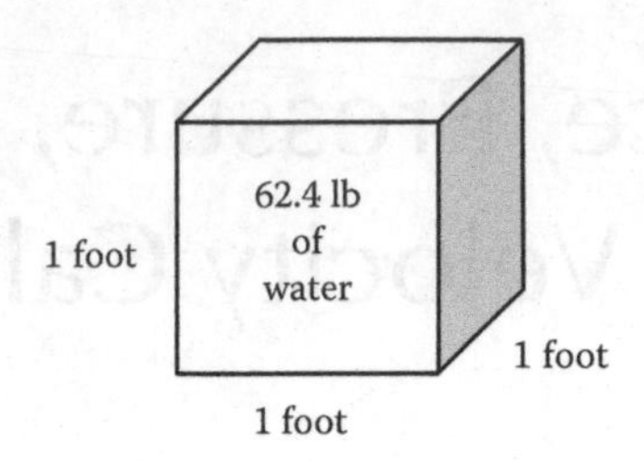

FIGURE 12.1 One cubic foot of water weighs 62.4 pounds.

If we use the bottom of the container as our reference point, the head would be 1 foot. From this we can see that 1 foot of head is equal to 0.433 psi. Figure 12.2 illustrates some other important relationships between pressure and head.

Note: Force acts in a particular direction. Water in a tank exerts force down on the bottom and out of the sides. Pressure, however, acts in all directions. A marble at a water depth of one foot would have 0.433 psi of pressure acting inward on all sides.

Note: Water and wastewater unit system pressures are measured in psi but centrifugal pumps are rated in feet of total dynamic head (TDH).

Water and wastewater operators must be able to convert from one pressure unit to the other. Thus, using the preceding information, we can develop Equations 12.1 and 12.2 for calculating pressure and head:

$$\text{Pressure (psi)} = 0.433 \times \text{Head (ft)} \tag{12.1}$$

$$\text{Head (ft)} = 2.31 \times \text{Pressure (psi)} \tag{12.2}$$

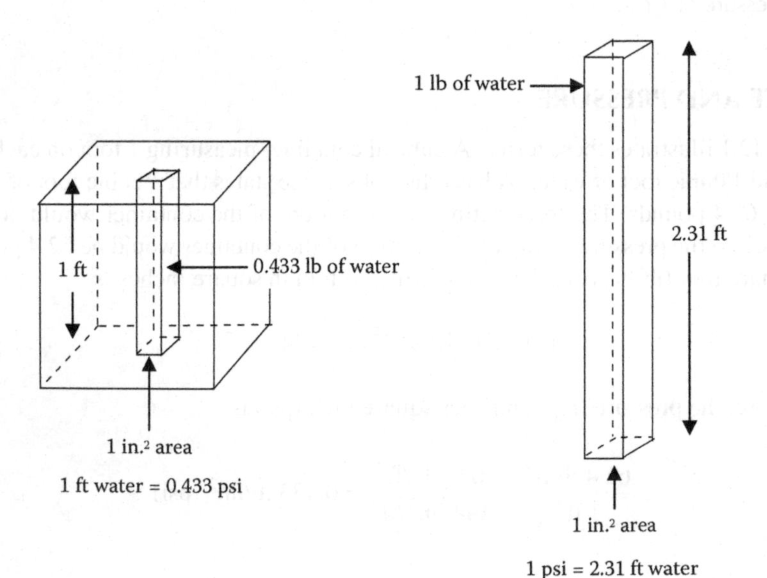

FIGURE 12.2 The relationship between pressure and head.

HEAD

Head is the vertical distance the water must be lifted from the supply tank or unit process to the discharge. If the pressure (psi) is known, the height of the water column can be determined by multiplying the psi by 2.31:

$$\text{psi} \times 2.31 = \text{Feet of head} \quad (12.3)$$

■ EXAMPLE 12.1

Problem: A pressure gauge at the bottom of a storage tank reads 35 psi. What is the water level in the tank?

Solution: Convert psi to feet of head.

$$35 \text{ psi} \times 2.31 = 80.9 \text{ ft of water above the gauge}$$

Note that if the height of a column of water is known, the pressure it exerts can be determined by dividing the feet of head by 2.31:

$$\text{Feet of head} \div 2.31 = \text{psi}$$

■ EXAMPLE 12.2

Problem: The reservoir level is 120 ft about the pump discharge. What is the discharge pressure on the pump?

Solution: Convert feet of head to psi.

$$120 \text{ ft} \div 2.31 = 51.9 \text{ psi}$$

■ EXAMPLE 12.3

Problem: A pump is installed at 5410 feet above sea level. The overflow of the reservoir is at 5530 feet above sea level. What is the discharge pressure on the pump in psi?

Solution: Find the difference in elevation:

$$5530 \text{ ft} - 5410 = 120 \text{ ft of head}$$

Convert feet of head to psi:

$$120 \text{ ft} \div 2.31 = 51.9 \text{ psi}$$

■ EXAMPLE 12.4

Problem: A discharge pressure gauge on a pump reads 74 psi when the pump is running. The pressure gauge at the top of a hill 40 ft above the pump reads 42 psi. What is the friction loss in the pipe in feet of head?

Solution: Find the difference in the pressures:

$$74 \text{ psi} - 42 \text{ psi} = 32 \text{ psi}$$

Convert psi to feet of head:

$$32 \text{ psi} \times 2.31 = 73.9 \text{ ft of head}$$

Subtract the difference in elevation to find the friction loss:

$$73.9 \text{ ft} - 40 \text{ ft} = 33.9 \text{ ft of head}$$

The total head includes the vertical distance the liquid must be lifted (static head), the loss to friction (friction head), and the energy required to maintain the desired velocity (velocity head):

$$\text{Total head} = \text{Static head} + \text{Friction head} + \text{Velocity head} \quad (12.4)$$

Static Head

Static head is the actual vertical distance the liquid must be lifted:

$$\text{Static head} = \text{Discharge elevation} - \text{Supply elevation} \quad (12.5)$$

■ Example 12.5

Problem: A supply tank is located at elevation 108 ft. The discharge point is at elevation 205 ft. What is the static head in feet?

Solution:

$$\text{Static head} = 205 \text{ ft} - 108 \text{ ft} = 97 \text{ ft}$$

Friction Head

Friction head is the equivalent distance of the energy that must be supplied to overcome friction. Engineering references include tables showing the equivalent vertical distance for various sizes and types of pipes, fittings, and valves. The total friction head is the sum of the equivalent vertical distances for each component:

$$\text{Friction head (ft)} = \text{Energy losses due to friction} \quad (12.6)$$

Velocity Head

Velocity head is the equivalent distance of the energy consumed in achieving and maintaining the desired velocity in the system:

$$\text{Velocity head (ft)} = \text{Energy losses due to maintaining velocity} \quad (12.7)$$

Total Dynamic Head (Total System Head)

$$\text{Total head} = \text{Static head} + \text{Friction head} + \text{Velocity head} \quad (12.8)$$

Pressure and Head

The pressure exerted by water or wastewater is directly proportional to its depth or head in the pipe, tank, or channel. If the pressure is known, the equivalent head can be calculated:

$$\text{Head (ft)} = \text{Pressure (psi)} \times 2.31 \text{ ft/psi} \quad (12.9)$$

■ Example 12.6

Problem: The pressure gauge on the discharge line from an influent pump reads 75.3 psi. What is the equivalent head in feet?

Solution:

$$\text{Head} = 75.3 \times 2.31 \text{ ft/psi} = 173.9 \text{ ft}$$

Head and Pressure

If the head is known, the equivalent pressure can be calculated by

$$\text{Pressure (psi)} = \text{Head (ft)} \div 2.31 \text{ ft/psi} \quad (12.10)$$

■ Example 12.7

Problem: A tank is 15 ft deep. What is the pressure in psi at the bottom of the tank when it is filled with wastewater?

Solution:

$$\text{Pressure (psi)} = 15 \text{ ft} \div 2.31 \text{ ft/psi} = 6.49 \text{ psi}$$

Before we look at a few example problems dealing with force, pressure, and head, it is important to summarize the key points related to force, pressure, and head:

1. By definition, water weighs 62.4 pounds per cubic foot.
2. The surface of any one side of a 1-ft cube contains 144 square inches (12 in. × 12 in. = 144 in.^{2}). Therefore, the cube contains 144 columns of water 1 foot tall and 1 inch square.
3. The weight of each of these pieces can be determined by dividing the weight of the water in the cube by the number of square inches.

$$\text{Weight} = 62.4 \text{ lb} \div 144 \text{ in.}^2 = 0.433 \text{ lb/in.}^2 \text{ (psi)}$$

4. Because this is the weight of one column of water 1 foot tall, the true expression would be 0.433 pounds per square inch per foot of head, or 0.433 psi/ft.

Note: 1 ft of head = 0.433 psi.

In addition to remembering the important parameter 1 foot of head = 0.433 psi, it is important to understand the relationship between pressure and feet of head—in other words, how many feet of head 1 psi represents. This is determined by dividing 1 by 0.433:

$$\text{Feet of head} = 1 \text{ ft} \div 0.433 \text{ psi} = 2.31 \text{ ft/psi}$$

If a pressure gauge reads 12 psi, the height of the water necessary to represent this pressure would be 12 psi × 2.31 ft/psi = 27.7 feet.

Note: Both the above conversions are commonly used in water/wastewater treatment calculations; however, the most accurate conversion is 1 ft = 0.433 psi. This is the conversion we use throughout this text.

VELOCITY

The velocity of the water moving through a pipe can be determined if the flow in cubic feet per second (cfs) and the diameter of the pipe (inches) are known. The area of the pipe must be calculated in square feet (ft^2) and the flow is then divided by the area.

$$\text{Velocity (fps)} = \frac{\text{Flow (cfs)}}{\text{Area (ft}^2\text{)}} \tag{12.11}$$

or

$$\text{Velocity (fps)} = \frac{\text{Distance}}{\text{Time}} \tag{12.12}$$

Velocity is expressed in units such as ft/sec, miles/hour, ft/min, etc. The time unit of velocity can vary, as long as it is the same within each problem.

■ EXAMPLE 12.8

Problem: A 24-in. pipe carries a flow of 12 cfs. What is the velocity in the pipe?

Solution: Change diameter in inches to feet:

$$24 \text{ in.} \div 12 \text{ in./ft} = 2 \text{ ft; radius} = 1 \text{ ft}$$

Find the area of the pipe in square feet:

$$\text{Area} = \pi \times r^2 = 3.14 \times (1 \text{ ft})^2 = 3.14 \text{ ft}^2$$

Find the velocity in fps:

$$12 \text{ cfs} \div 3.14 \text{ ft}^2 = 3.8 \text{ fps}$$

Note: The flow through a pipe (cfs) can be determined if the velocity and pipe diameter are known. The area of the pipe must be calculated in square feet and then multiplied by the velocity (fps).

FORCE, PRESSURE, HEAD, AND VELOCITY EXAMPLES

■ Example 12.9

Problem: Convert 40 psi to feet of head.

Solution:

$$40\text{ psi} \times \frac{1\text{ ft}}{0.433\text{ psi}} = 92.4\text{ ft}$$

■ Example 12.10

Problem: Convert 40 feet of head to psi.

Solution:

$$40\text{ ft} \times \frac{0.433\text{ psi}}{1\text{ ft}} = 17.32\text{ psi}$$

As the above examples demonstrate, when we convert psi to feet we divide by 0.433, and when we convert feet to psi we multiply by 0.433. This information can be most helpful in clearing up any confusion as to whether to multiply or divide. There is another way to think about it, though—one that may be more beneficial and easier for many operators to use. Notice that the relationship between psi and feet is almost two to one. It takes slightly more than 2 ft to make 1 psi. Therefore, when looking at a problem where the information provided is in pressure and the result should be in feet, the answer will be at least twice as large as the starting number. For example, if the pressure is 25 psi, we intuitively know that the head is over 50 ft. Therefore, we must divide by 0.433 to obtain the correct answer.

■ Example 12.11

Problem: Convert a pressure of 45 psi to feet of head.

Solution:

$$45\text{ psi} \times \frac{1\text{ ft}}{0.433\text{ psi}} = 104\text{ ft}$$

■ Example 12.12

Problem: Convert a pressure of 15 psi to feet of head.

Solution:

$$15\text{ psi} \times \frac{1\text{ ft}}{0.433\text{ psi}} = 34.6\text{ ft}$$

■ Example 12.13

Problem: Between the top of a reservoir and the watering point, the elevation is 125 ft. What will the static pressure be at the watering point?

Solution:

$$125\text{ ft} \times \frac{0.433\text{ psi}}{1\text{ ft}} = 289\text{ ft}$$

■ Example 12.14

Problem: Find the pressure in psi in a tank 12 ft deep at a point 5 ft below the water surface.

Solution:

$$\text{Pressure} = 0.433 \times 5\text{ ft} = 2.17\text{ psi}$$

■ Example 12.15

Problem: A pressure gauge at the bottom of a tank reads 12.2 psi. How deep is the water in the tank?

Solution:

$$\text{Head (ft)} = 2.31\text{ ft/psi} \times 12.2\text{ psi} = 28.2\text{ ft}$$

■ Example 12.16

Problem: What is the static pressure 4 miles beneath the ocean surface?

Solution: Change miles to feet and then to pounds per square inch.

$$5280\text{ ft/mile} \times 4\text{ miles} = 21{,}120\text{ ft}$$

$$21{,}120\text{ ft} \div 2.31\text{ ft/psi} = 9143\text{ psi}$$

■ Example 12.17

Problem: A 150-ft diameter cylindrical tank contains 2 MG of water. What is the water depth?

Solution:

1. Change MG to cubic feet.

$$(2{,}000{,}000\text{ gal})/(7.48\text{ gal/ft}^3) = 267{,}380\text{ ft}^3$$

2. Using volume, solve for depth.

$$\text{Volume} = 0.785 \times (\text{Diameter})^2 \times \text{Depth}$$

$$267{,}380\text{ ft}^3 = 0.785 \times (150\text{ ft})^2 \times \text{Depth}$$

$$\text{Depth} = 15.1\text{ ft}$$

■ Example 12.18

Problem: The pressure in a pipe is 70 psi. What is the pressure in feet of water? What is the pressure in pounds per square foot?

Solution:

1. Convert pressure to feet of water.

$$\text{Pressure} = 70 \text{ psi} \times 2.31 \text{ ft/psi} = 161.7 \text{ ft of water}$$

2. Convert psi to psf.

$$\text{Pressure} = 70 \text{ psi} \times 144 \text{ in.}^2/\text{ft}^2 = 10{,}080 \text{ psf}$$

■ Example 12.19

Problem: The pressure in a pipeline is 6476 psf. What is the head on the pipe?

Solution:

$$\text{Head on pipe} = \text{Feet of pressure}$$

$$\text{Pressure} = \text{Weight} \times \text{Height}$$

$$6476 \text{ psf} = 62.4 \text{ lb/ft}^3 \times \text{Height}$$

$$\text{Height} = 104 \text{ ft}$$

■ Example 12.20

Problem: A 12-in. pipe carries water at a velocity of 6 fps. What is the flow in cfs?

Solution:

1. Change inches to feet.

$$12 \text{ in.} \div 12 \text{ in./ft} = 1 \text{ ft; radius} = 0.5 \text{ ft}$$

2. Find the area of the pipe in square feet.

$$\text{Area} = \pi \times r^2 = 3.14 \times (0.5 \text{ ft})^2 = 0.785 \text{ ft}^2$$

3. Find the flow in cfs.

$$\text{Flow} = 6 \text{ fps} \times 0.785 \text{ ft}^2 = 4.71 \text{ cfs}$$

■ Example 12.21

Problem: A 12-in. pipe carries 1400 gpm at a velocity of 4 fps and reduces to an 8-in. pipe. What is the velocity in the 6-in. pipe?

Solution:

1. Convert flow to cfs.

$$1400 \text{ gpm} \div 449 \text{ gpm/cfs} = 3.12 \text{ cfs}$$

2. Change inches to feet.

$$8 \text{ in.} \div 12 \text{ in./ft} = 0.666 \text{ ft}$$

3. Find the area of the pipe in square feet.

$$\text{Area} = \pi \times r^2 = 3.14 \times (0.333 \text{ ft})^2 = 0.349 \text{ ft}^2$$

4. Find the velocity in fps.

$$\text{Velocity} = 3.12 \text{ cfs} \div 0.349 \text{ ft}^2 = 9 \text{ fps}$$

■ Example 12.22

Problem: A rocket traveled 4200 ft in 6 seconds. What was the velocity of the rocket in feet per second?

Solution:

$$\text{Velocity} = \frac{\text{Distance}}{\text{Time}} = \frac{4200 \text{ ft}}{6 \text{ sec}} = 700 \text{ fps}$$

■ Example 12.23

Problem: A bobber is placed in a stream and travels 520 ft in 2.5 minutes. What is the velocity of the water flowing in the stream in feet per minute?

Solution:

$$\text{Velocity} = \frac{\text{Distance}}{\text{Time}} = \frac{520 \text{ ft}}{2.5 \text{ min}} = 208 \text{ ft/min}$$

■ Example 12.24

Problem: A cork is placed in a channel and travels 390 ft in 2 minutes. What is the velocity of the water in the channel in feet per minute?

Solution:

$$\text{Velocity} = \frac{\text{Distance}}{\text{Time}} = \frac{390 \text{ ft}}{2 \text{ min}} = 195 \text{ ft/min}$$

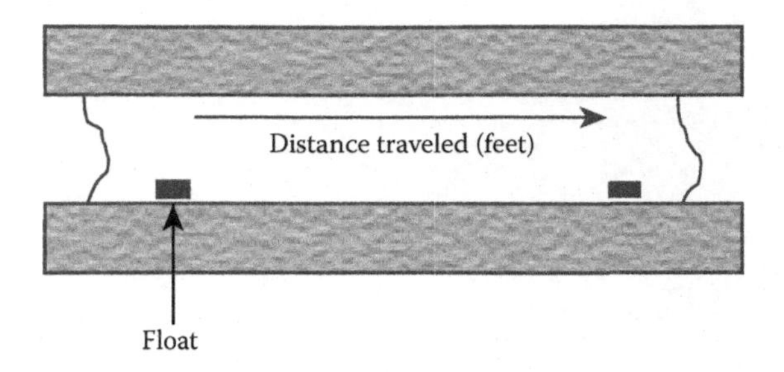

FIGURE 12.3 Illustration for Example 12.25.

■ Example 12.25

Problem: A float travels 320 ft in a channel (see Figure 12.3) in 2 minutes and 20 seconds. What is the velocity in the channel in feet per second?

Solution:

$$2\text{ min, }20\text{ sec} = (2 \times 60) + 20 = 140\text{ sec}$$

$$\text{Velocity} = 320\text{ ft} \div 140\text{ sec} = 2.29\text{ ft/sec (fps)}$$

■ Example 12.26

Problem: The distance between manhole #1 and manhole #2 is 110 ft. A fishing bobber is dropped into manhole #1 and enters manhole #2 in 30 seconds. What is the velocity of the wastewater in the sewer in feet per minute?

Solution:

$$30\text{ sec} \div 60\text{ sec/min} = 0.5\text{ min}$$

$$\text{Velocity} = 110\text{ ft} \div 0.5\text{ min} = 220\text{ ft/min}$$

■ Example 12.27

Problem: A float is placed in a channel. It takes 2.5 minutes to travel 400 ft. What is the flow velocity in feet per minute in the channel? (Assume that the float is traveling at the average velocity of the water.)

Solution:

$$\text{Velocity} = \frac{\text{Distance}}{\text{Time}} = \frac{400\text{ ft}}{2.5\text{ min}} = 160\text{ ft/min}$$

■ Example 12.28

Problem: A cork placed in a channel travels 40 feet in 20 seconds. What is the velocity of the cork in feet per second?

Solution:

$$\text{Velocity} = 40\text{ ft} \div 20\text{ sec} = 2\text{ ft/sec (fps)}$$

13 Mass Balance and Measuring Plant Performance

The simplest way to express the fundamental engineering principle of *mass balance* is to say, "Everything has to go somewhere." More precisely, the *law of conservation of mass* says that when chemical reactions take place, matter is neither created nor destroyed. What this important concept allows us to do is track materials (e.g., pollutants, microorganisms, chemicals) from one place to another. The concept of mass balance plays an important role in treatment plant operations (especially wastewater treatment) where we assume that a balance exists between the material entering and leaving the treatment plant or a treatment process: "What comes in must equal what goes out." The concept is very helpful in evaluating biological systems, sampling and testing procedures, and many other unit processes within the treatment system. In the following sections, we illustrate how the mass balance concept is used to determine the quantity of solids entering and leaving settling tanks and mass balance using biochemical oxygen demand (BOD) removal.

MASS BALANCE FOR SETTLING TANKS

The mass balance for the settling tank calculates the quantity of solids entering and leaving the unit.

> ***Note:*** The two numbers—in (influent) and out (effluent)—must be within 10 to 15% of each other to be considered acceptable. Larger discrepancies may indicate sampling errors, increasing solids levels in the unit, or undetected solids discharge in the tank effluent.

To get a better feel for how the mass balance for settling tanks procedure is formatted for actual use, refer to Figure 13.1 and the steps provided below, which are used in Example 13.1 to demonstrate how mass balance is actually used in wastewater operations.

Step 1. Solids in = Pounds of influent suspended solids
Step 2. Solids out = Pounds of effluent suspended solids
Step 3. Biosolids out = Pounds of biosolids pumped per day
Step 4. Balance = Solids in – (Solids out + Biosolids pumped)

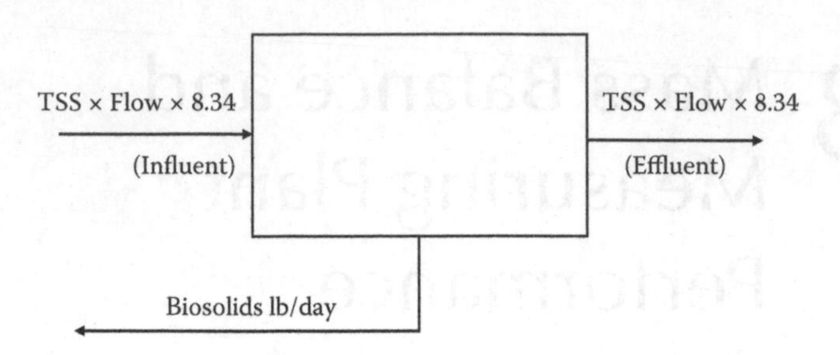

FIGURE 13.1 Mass balance for settling tanks.

■ Example 13.1

Problem: A settling tank receives a daily flow of 4.20 million gallons per day (MGD). The influent contains 252 mg/L suspended solids, and the unit effluent contains 140 mg/L suspended solids. The biosolids pump operates 10 min/hr and removes biosolids at the rate of 40 gallons per minute (gpm). The biosolids are comprised of 4.2% solids. Determine if the mass balance for solids removal is within the acceptable 10 to 15% range.

Solution:

Step 1. Solids in (influent) = 252 mg/L × 4.20 MGD × 8.34 lb/gal = 8827 lb/day

Step 2. Solids out (effluent) = 140 mg/L × 4.20 MGD × 8.34 lb/gal = 4904 lb/day

Step 3. Calculate biosolids out:

$$\text{Biosolids out} = 10\text{ min/hr} \times 24\text{ hr/day} \times 40\text{ gpm} \times 8.34\text{ lb/gal} \times 0.042 = 3363\text{ lb/day}$$

Step 4. Mass balance = 8827 lb/day – (4904 lb/day + 3363 lb/day) = 560 lb, or 6.3%

MASS BALANCE USING BOD REMOVAL

The amount of BOD removed by a treatment process is directly related to the quantity of solids the process will generate. Because the actual amount of solids generated will vary with operational conditions and design, exact figures must be determined on a case-by-case basis; however, research has produced general conversion rates for many of the common treatment processes. These values are given in Table 13.1 and can be used if plant-specific information is unavailable. Using these factors, the mass balance procedure determines the amount of solids the process is anticipated to produce. This is compared with the actual biosolids production to determine the accuracy of the sampling, the potential for solids buildup in the system, or unrecorded solids discharges (see Figure 13.2).

1. BOD_{in} = Influent BOD × Flow × 8.34 lb/gal
2. BOD_{out} = Effluent BOD × Flow × 8.34 lb/gal
3. BOD removed (lb) = $BOD_{in} - BOD_{out}$

TABLE 13.1
General Conversion Rates

Process Type	Conversion Factor (lb Solids/lb BOD Removal)
Primary treatment	1.70
Trickling filters	1.00
Rotating biological contactors	1.00
Activated biosolids with primary	0.70
Activated biosolids without primary	
Conventional	0.85
Extended air	0.65
Contact stabilization	1.00
Step feed	0.85
Oxidation ditch	0.65

4. Solids generated (lb) = BOD removed (lb) × Factor
5. Solids removed (lb/day) = Sludge pumped (gpd) × % Solids × 8.34 lb/gal
6. Effluent solids (mg/L) = Flow (MGD) × 8.34 lb/gal

■ Example 13.2

Problem: A conventional activated biosolids system with primary treatment is operating at the levels listed below. Does the mass balance for the activated biosolids system indicate that a problem exists?

Plant influent BOD = 250 mg/L
Primary effluent BOD = 166 mg/L
Activated biosolids system effluent BOD = 25 mg/L
Activated biosolids system effluent total suspended solids (TSS) = 19 mg/L
Plant flow = 11.40 MGD
Waste concentration = 6795 mg/L
Waste flow = 0.15 MGD

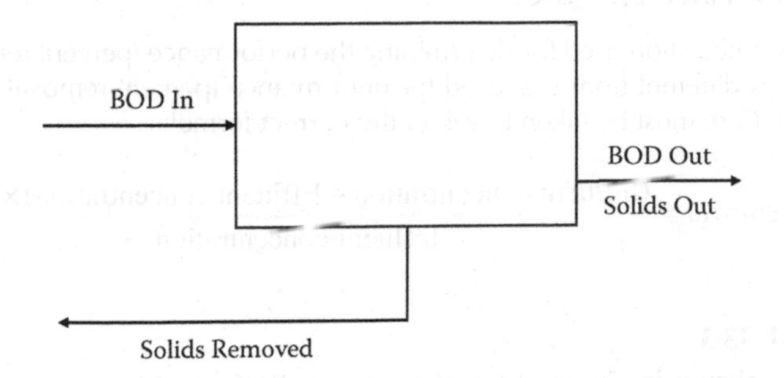

FIGURE 13.2 Comparison of BOD_{in} and BOD_{out}.

Solution:

BOD_{in} = 166 mg/L × 11.40 MGD × 8.34 lb/gal = 15,783 lb/day
BOD_{out} = 25 mg/L × 11.40 MGD × 8.34 lb/gal = 2377 lb/day
BOD removed = 15,783 lb/day – 2377 lb/day = 13,406 lb/day
Solids produced (lb/day) = 13,406 lb/day × 0.7 lb solids/lb BOD = 9384 lb/day
Solids removed (lb/day) = 6795 mg/L × 0.15 MGD × 8.34 lb/gal = 8501 lb/day
Difference = 9384 lb/day – 8501 lb/day = 883 lb/day, or 9.4%

These results are within the acceptable range.

Note: We have demonstrated two ways in which mass balance can be used; however, it is important to note that the mass balance concept can be used for all aspects of wastewater and solids treatment. In each case, the calculations must take into account all of the sources of material entering the process and all of the methods available for removal of solids.

MEASURING PLANT PERFORMANCE

To evaluate how well a plant or unit process is performing, performance efficiency or percent removal is used. The results obtained can be compared with those listed in the plant's operations and maintenance (O&M) manual to determine if the facility is performing as expected. In this section, sample calculations often used to measure plant performance or efficiency are presented. The *efficiency* of a unit process is its effectiveness in removing various constituents from the wastewater or water. Suspended solids and BOD removal are therefore the most common calculations of unit process efficiency. In wastewater treatment, the efficiency of a sedimentation basin may be affected by such factors as the types of solids in the wastewater, the temperature of the wastewater, and the age of the solids. Typical removal efficiencies for a primary sedimentation basin are as follows:

Settleable solids, 90–99%
Suspended solids, 40–60%
Total solids, 10–15%
BOD, 20–50%

PLANT PERFORMANCE/EFFICIENCY

Note: The calculation used for determining the performance (percent removal) for a digester is different from that used for performance (percent removal) for other processes. Care must be taken to select the correct formula:

$$\%\ \text{Removal} = \frac{(\text{Influent concentration} - \text{Effluent concentration}) \times 100}{\text{Influent concentration}} \quad (13.1)$$

■ EXAMPLE 13.3

Problem: As shown in Figure 13.3, the influent BOD_5 is 247 mg/L, and the plant effluent BOD is 17 mg/L. What is the percent removal?

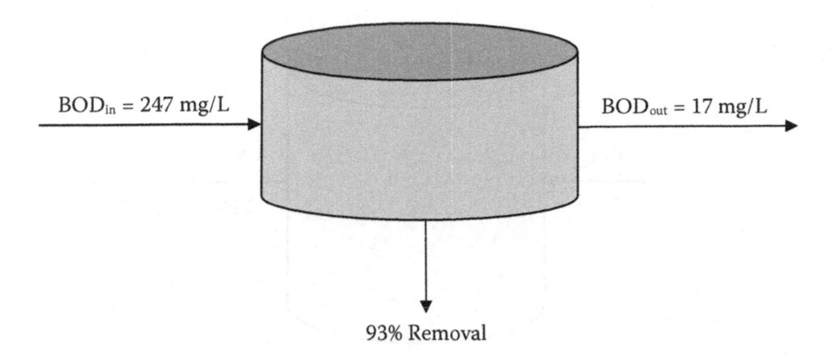

FIGURE 13.3 Illustration for Example 13.3.

Solution:

$$\% \text{ Removal} = \frac{(247 \text{ mg/L} - 17 \text{ mg/L}) \times 100}{247 \text{ mg/L}} = 93\%$$

Unit Process Performance and Efficiency

Equation 13.1 is also used to determine unit process efficiency. The concentration entering the unit and the concentration leaving the unit (e.g., primary, secondary) are used to determine the unit performance.

■ Example 13.4

Problem: As shown in Figure 13.4, the primary influent BOD is 235 mg/L, and the primary effluent BOD is 169 mg/L. What is the percent removal?

Solution:

$$\% \text{ Removal} = \frac{(235 \text{ mg/L} - 169 \text{ mg/L}) \times 100}{235 \text{ mg/L}} = 28\%$$

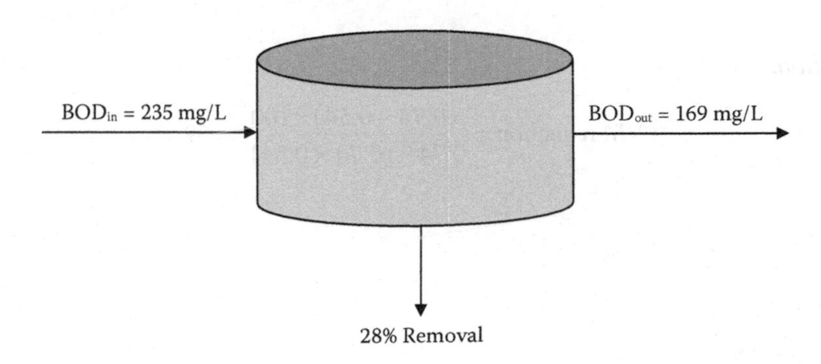

FIGURE 13.4 Illustration for Example 13.4.

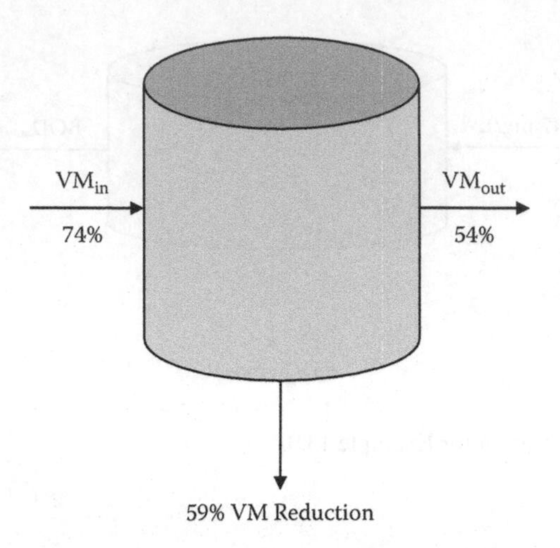

FIGURE 13.5 Illustration for Example 13.5.

Percent Volatile Matter Reduction in Sludge

The calculation used to determine percent volatile matter reduction is more complicated because of the changes occurring during biosolids digestion:

$$\%\text{VM reduction} = \frac{(\%\text{VM}_{\text{in}} - \%\text{VM}_{\text{out}}) \times 100}{\%\text{VM}_{\text{in}} - (\%\text{VM}_{\text{in}} \times \%\text{VM}_{\text{out}})} \tag{13.2}$$

■ Example 13.5

Problem: Using the digester data provided below, determine the percent volatile matter reduction for the digester (see Figure 13.5).

Raw biosolids volatile matter = 74%
Digested biosolids volatile matter = 54%

Solution:

$$\%\text{VM reduction} = \frac{(0.74 - 0.54) \times 100}{0.74 - (0.74 \times 0.54)} = 59\%$$

14 Pumping Calculations

Pumping facilities and appurtenances are required wherever gravity cannot be used to supply water to the distribution system under sufficient pressure to meet all service demands. For wastewater, pumps are used to lift or elevate the liquid from a lower elevation to an adequate height at which it can flow by gravity or overcome hydrostatic head. There are many pumping applications at a wastewater treatment facility. These applications include pumping of (1) raw or treated wastewater, (2) grit, (3) grease and floating solids, (4) dilute or well-thickened raw biosolids or digested biosolids (biosolids or supernatant return), and (5) chemical solutions. Pumps and lift stations are used extensively in the collection system. Each of the various pumping applications is unique and requires specific design and pump selection considerations.

With few exceptions, the pumps used in water and wastewater treatment are the same. Because the pump is so perfectly suited to the tasks it performs, and because the principles that make the pump work are physically fundamental, the idea that any new device would ever replace the pump is difficult to imagine. The pump is the workhorse of water/wastewater operations. Simply, pumps use energy to keep water and wastewater moving. To operate a pump efficiently, the operator and/or maintenance operator must be familiar with several basic principles of hydraulics. In addition, to operate various unit processes, in both water and wastewater operations at optimum levels, operators should know how to perform basic pumping calculations.

BASIC WATER HYDRAULICS CALCULATIONS

Weight of Water

Because water must be stored or kept moving in water supplies and wastewater must be collected, processed, and discharged (outfalled) to its receiving body, we must consider some basic relationships with regard to the weight of water. One cubic foot of water weighs 62.4 pounds and contains 7.48 gallons. One cubic inch of water weighs 0.0362 pounds. Water 1 foot deep will exert a pressure of 0.43 pounds per square inch (psi) on the bottom area (12 in. × 0.062 lb/in.^{3}). A column of water 2 feet high exerts 0.86 psi, one 10 feet high exerts 4.3 psi, and one 52 feet high exerts

$$52 \text{ ft} \times 0.43 \text{ psi/ft} = 22.36 \text{ psi}$$

A column of water 2.31 feet high will exert 1.0 psi. To produce a pressure of 40 psi requires a water column calculated as follows:

$$40 \text{ psi} \times 2.31 \text{ ft/psi} = 92.4 \text{ ft}$$

The term *head* is used to designate water pressure in terms of the height of a column of water in feet; for example, a 10-foot column of water exerts 4.3 psi. This can be referred to as 4.3 psi pressure or 10 feet of head. If the static pressure in a pipe leading from an elevated water storage tank is 37 psi, what is the elevation of the water above the pressure gauge? Remembering that 1 psi = 2.31 and that the pressure at the gauge is 37 psi, then

$$37 \text{ psi} \times 2.31 \text{ ft/psi} = 85.5 \text{ ft}$$

Weight of Water Related to the Weight of Air

The theoretical atmospheric pressure at sea level (14.7 psi) will support a column of water 34 ft high:

$$14.7 \text{ psi} \times 2.31 \text{ ft/psi} = 34 \text{ ft}$$

At an elevation of 1 mile above sea level, where the atmospheric pressure is 12 psi, the column of water would be only 28 ft high:

$$12 \text{ psi} \times 2.31 \text{ ft/psi} = 28 \text{ ft}$$

If a tube is placed in a body of water at sea level (a glass, a bucket, a water storage reservoir, a lake or pool), water will rise in the tube to the same height as the water outside the tube. The atmospheric pressure of 14.7 psi will push down equally on the water surface inside and outside the tube. However, if the top of the tube is tightly capped and all of the air is removed from the sealed tube above the water surface, forming a *perfect vacuum*, the pressure on the water surface inside the tube will be 0 psi. The atmospheric pressure of 14.7 psi on the outside of the tube will push the water up into the tube until the weight of the water exerts the same 14.7-psi pressure at a point in the tube even with the water surface outside the tube. The water will rise 14.7 psi × 2.31 ft/psi = 34 ft. In practice, it is impossible to create a perfect vacuum, so the water will rise somewhat less than 34 ft; the distance it rises depends on the amount of vacuum created.

■ Example 14.1

Problem: If enough air was removed from the tube to produce an air pressure of 9.7 psi above the water in the tube, how far will the water rise in the tube?

Solution: To maintain the 14.7 psi at the outside water surface level, the water in the tube must produce a pressure of 14.7 psi – 9.7 = 5.0 psi. The height of the column of water that will produce 5.0 psi is

$$5.0 \text{ psi} \times 2.31 \text{ ft/psi} = 11.5 \text{ ft}$$

Water at Rest

Stevin's law states: "The pressure at any point in a fluid at rest depends on the distance measured vertically to the free surface and the density of the fluid." Stated as a formula, this becomes

$$p = w \times h \tag{14.1}$$

where

p = Pressure in pounds per square foot (lb/ft² or psf).
w = Density in pounds per cubic foot (lb/ft³).
h = Vertical distance in feet (ft).

■ Example 14.2

Problem: What is the pressure at a point 15 ft below the surface of a reservoir?

Solution: To calculate this, we must know that the density (w) of water is 62.4 lb/ft³:

$$p = w \times h = 62.4 \text{ lb/ft}^3 \times 15 \text{ ft} = 936 \text{ lb/ft}^2 \text{ (psf)}$$

Water/wastewater operators generally measure pressure in pounds per square inch rather than pounds per square foot; to convert, divide by 144 in.²/ft² (12 in. × 12 in. = 144 in.²):

$$936 \text{ lb/ft}^2 \div 144 \text{ in.}^2\text{/ft}^2 = 6.5 \text{ lb/in.}^2 \text{ (psi)}$$

Gauge Pressure

We defined head as the height a column of water would rise due to the pressure at its base. We demonstrated that a perfect vacuum plus atmospheric pressure of 14.7 psi would lift the water 34 ft. If we now open the top of the sealed tube to the atmosphere and enclose the reservoir, then increase the pressure in the reservoir, the water will again rise in the tube. Because atmospheric pressure is essentially universal, we usually ignore the first 14.7 psi of actual pressure measurements and measure only the difference between the water pressure and the atmospheric pressure; we call this *gauge pressure*.

■ Example 14.3

Problem: Water in an open reservoir is subjected to the 14.7 psi of atmospheric pressure, but subtracting this 14.7 psi leaves a gauge pressure of 0 psi. This shows that the water would rise 0 feet above the reservoir surface. If the gauge pressure in a water main is 100 psi, how far would the water rise in a tube connected to the main?

Solution:

$$100 \text{ psi} \times 2.31 \text{ ft/psi} = 231 \text{ ft}$$

Water in Motion

The study of water in motion is much more complicated than that of water at rest. It is important to have an understanding of these principles because the water/wastewater in a treatment plant or distribution/collection system is nearly always in motion (much of this motion is the result of pumping, of course).

Discharge

Discharge is the quantity of water passing a given point in a pipe or channel during a given period of time. It can be calculated by the following formula:

$$Q = A \times V \tag{14.2}$$

where

Q = Flow, or discharge, in cubic feet per second (cfs).
A = Cross-sectional area of the pipe or channel in square feet (ft^2).
V = Water velocity in feet per second (fps or ft/sec).

Discharge can be converted from cfs to other units such as gallons per minute (gpm) or million gallons per day (MGD) by using appropriate conversion factors.

■ Example 14.4

Problem: A 12-in.-diameter pipe has water flowing through it at 10 fps. What is the discharge in (a) cfs, (b) gpm, and (c) MGD?

Solution: Before we can use the basic formula, we must determine the area (A) of the pipe. The formula for the area of a circle is

$$\text{Area} = \pi \times \frac{D^2}{4} = \pi \times r^2$$

where

π = Constant value 3.14159, or simply 3.14.
D = Diameter of the circle (ft).
r = Radius of the circle (ft).

Therefore, the area of the pipe is

$$\text{Area} = \pi \times \frac{D^2}{4} = \pi \times \frac{(1\text{ ft})^2}{4} = 3.14 \times 1/4\text{ ft} = 0.785\text{ ft}^2$$

(a) Now we can determine the discharge in cfs:

$$Q = V \times A = 10\text{ ft/sec} \times 0.785\text{ ft}^2 = 7.85\text{ ft}^3\text{/sec (cfs)}$$

(b) We know that 1 cfs is 449 gpm, so

$$7.85\text{ cfs} \times 449\text{ gpm/cfs} = 3525\text{ gpm}$$

(c) 1 million gallons per day is 1.55 cfs, so

$$7.85 \text{ cfs} \div 1.55 \text{ cfs/MGD} = 5.06 \text{ MGD}$$

Law of Continuity

The law of continuity states that the discharge at each point in a pipe or channel is the same as the discharge at any other point (if water does not leave or enter the pipe or channel). That is, under the assumption of steady-state flow, the flow that enters the pipe or channel is the same flow that exits the pipe or channel. In equation form, this becomes

$$Q_1 = Q_2, \text{ or } A_1 \times V_1 = A_2 \times V_2 \tag{14.3}$$

■ Example 14.5

Problem: A pipe 12 inches in diameter is connected to a 6-in.-diameter pipe. The velocity of the water in the 12-in. pipe is 3 fps. What is the velocity in the 6-in. pipe?

Solution: To use the equation $A_1 \times V_1 = A_2 \times V_2$, we need to determine the area of each pipe:

$$\text{Area of 12-in. pipe} = \pi \times \frac{D^2}{4} = 3.14 \times \frac{(1 \text{ ft})^2}{4} = 0.785 \text{ ft}^2$$

$$\text{Area of 6-in. pipe} = \pi \times \frac{D^2}{4} = 3.14 \times \frac{(0.5 \text{ ft})^2}{4} = 0.196 \text{ ft}^2$$

The continuity equation now becomes

$$A_1 \times V_1 = A_2 \times V_2$$

$$0.785 \text{ ft}^2 \times 3 \text{ ft/sec} = 0.196 \text{ ft}^2 \times V_2$$

Solving for V_2,

$$0.785 \text{ ft}^2 \times 3 \text{ ft/sec} = 0.196 \text{ ft}^2 \times V_2$$

$$\frac{0.785 \text{ ft}^2 \times 3 \text{ ft/sec}}{0.196 \text{ ft}^2} = V_2$$

$$12 \text{ ft/sec} = V_2$$

Pipe Friction

The flow of water in pipes is caused by the pressure applied behind it either by gravity or by hydraulic machines (pumps). The flow is retarded by the friction of the water against the inside of the pipe. The resistance of flow offered by this friction depends on the size (diameter) of the pipe, the roughness of the pipe wall, and the number and type of fittings (bends, valves, etc.) along the pipe. It also depends on the speed of the water through the pipe—the more water you try to pump through a pipe, the more pressure it will take to overcome the friction. The resistance can be expressed in terms of the additional pressure needed to push the water through the pipe, in either psi or feet of head. Because it is a reduction in pressure, it is often referred to as *friction loss* or *head loss*.

Friction loss increases as

- Flow rate increases.
- Pipe length increases.
- Pipe diameter decreases.
- Pipe is constricted.
- Pipe interior becomes rougher.
- Bends, fittings, and valves are added.

The actual calculation of friction loss is beyond the scope of this text. Many published tables give the friction loss in different types and diameters of pipe and standard fittings. What is more important here is recognition of the loss of pressure or head due to the friction of water flowing through a pipe. One of the factors in friction loss is the roughness of the pipe wall. A number called the *C* factor indicates pipe wall roughness; the higher the *C* factor, the smoother the pipe.

Note: The *C* factor is derived from the variable *C* in the Hazen–Williams equation for calculating water flow through a pipe.

Some of the roughness in the pipe will be due to the material; cast iron pipe will be rougher than plastic, for example. Additionally, the roughness will increase with corrosion of the pipe material and deposits of sediments in the pipe. New water pipes should have a *C* factor of 100 or more; older pipes can have *C* factors very much lower than this. To determine the *C* factor, we usually use published tables. In addition, when the friction losses for fittings are factored in, other published tables are available to make the proper determinations. It is standard practice to calculate the head loss from fittings by substituting the *equivalent length of pipe*, which is also available from published tables.

BASIC PUMPING CALCULATIONS

Certain computations used for determining various pumping parameters are important to the water/wastewater operator. In this section, we cover important basic pumping calculations.

Pumping Rates

Note: The rate of flow produced by a pump is expressed as the volume of water pumped during a given period.

The mathematical problems most often encountered by water/wastewater operators with regard to determining pumping rates are often solved by using Equations 14.4 and 14.5:

$$\text{Pumping rate (gpm)} = \text{Gallons} \div \text{Minutes} \tag{14.4}$$

$$\text{Pumping rate (gph)} = \text{Gallons} \div \text{Hours} \tag{14.5}$$

■ Example 14.6

Problem: The meter on the discharge side of the pump reads in hundreds of gallons. If the meter shows a reading of 110 at 2:00 p.m. and 320 at 2:30 p.m., what is the pumping rate expressed in gallons per minute (gpm)?

Solution: The problem asks for pumping rate in gallons per minute, so we use Equation 14.4:

$$\text{Pumping rate (gpm)} = \text{Gallons} \div \text{Minutes}$$

To solve this problem, we must first find the total gallons pumped (determined from the meter readings):

$$32{,}000 \text{ gal} - 11{,}000 \text{ gal} = 21{,}000 \text{ gal}$$

The volume was pumped between 2:00 p.m. and 2:30 p.m., for a total of 30 minutes. From this information, calculate the gallons-per-minute pumping rate:

$$\text{Pumping rate} = 21{,}000 \text{ gal} \div 30 \text{ min} = 700 \text{ gpm}$$

■ Example 14.7

Problem: During a 15-minute pumping test, 16,400 gal were pumped into an empty rectangular tank. What is the pumping rate in gallons per minute?

Solution: The problem asks for the pumping rate in gallons per minute, so again we use Equation 14.4:

$$\text{Pumping rate} = \text{Gallons} \div \text{Minutes} = 16{,}400 \text{ gal} \div 15 \text{ min} = 1093 \text{ gpm}$$

■ Example 14.8

Problem: A tank 50 ft in diameter is filled with water to a depth of 4 ft. To conduct a pumping test, the outlet valve to the tank is closed and the pump is allowed to discharge into the tank. After 80 minutes, the water level is 5.5 ft. What is the pumping rate in gallons per minute?

Solution: We must first determine the volume pumped in cubic feet:

$$\text{Volume pumped} = \text{Area of circle} \times \text{Depth} = 0.785 \times (50 \text{ ft})^2 \times 1.5 \text{ ft} = 2944 \text{ ft}^3$$

Now convert the cubic-foot volume to gallons:

$$2944 \text{ ft}^3 \times 7.48 \text{ gal/ft}^3 = 22{,}021 \text{ gal}$$

The pumping test was conducted over a period of 80 minutes. Using Equation 14.4, calculate the pumping rate in gallons per minute:

Pumping rate = Gallons ÷ Minutes = 22,021 gal ÷ 80 min = 275.3 gpm

Calculating Head Loss

Note: Pump head measurements are used to determine the amount of energy a pump can or must impart to the water; they are measured in feet.

One of the principle calculations in pumping problems is used to determine head loss. The following formula is used to calculate head loss:

$$H_f = K \times (V^2/2g) \tag{14.6}$$

where

H_f = Friction head.
K = Friction coefficient.
V = Velocity in pipe.
g = Gravity (32.17 ft/sec · sec)

Calculating Head

For centrifugal pumps and positive-displacement pumps, several other important formulas are used to determine head. In centrifugal pump calculations, conversion of the discharge pressure to discharge head is the norm. Positive-displacement pump calculations often leave given pressures in psi. In the following formulas, W expresses the specific weight of liquid in pounds per cubic foot. For water at 68°F, W is 62.4 lb/ft³. A water column 2.31 feet high exerts a pressure of 1 psi on 64°F water. Use the following formulas to convert discharge pressure in psig to head in feet:

- Centrifugal pump

$$\text{Head (ft)} = \frac{\text{Pressure (psig)} \times 2.31}{\text{Specific gravity}} \tag{14.7}$$

- Positive-displacement pump

$$\text{Head (ft)} = \frac{\text{Pressure (psig)} \times 144}{W} \tag{14.8}$$

Use the following formulas to convert head in feet to psig:

- Centrifugal pumps

$$\text{Pressure (psig)} = \frac{\text{Head (ft)} \times \text{Specific gravity}}{2.31} \tag{14.9}$$

- Positive-displacement pumps

$$\text{Pressure (psig)} = \frac{\text{Head (ft)} \times W}{144} \tag{14.10}$$

Calculating Horsepower and Efficiency

When considering work being done, we consider the "rate" at which work is being done. This is called *power* and is labeled as foot-pounds/second (ft-lb/sec). At some point in the past, it was determined that the ideal work animal, the horse, could move 550 pounds a distance of 1 foot in 1 second. Because large amounts of work are also to be considered, this unit became known as *horsepower.* When pushing a certain amount of water at a given pressure, the pump performs work. One horsepower equals 33,000 ft-lb/min. The two basic terms for horsepower are

- Hydraulic horsepower (whp)
- Brake horsepower (bhp)

Hydraulic Horsepower

One hydraulic horsepower (whp) equals the following:

- 550 ft-lb/sec
- 33,000 ft-lb/min
- 2545 British thermal units per hour (Btu/hr)
- 0.746 kW
- 1.014 metric hp

To calculate the hydraulic horsepower using flow in gpm and head in feet, use the following formula for centrifugal pumps:

$$\text{whp} = \frac{\text{Flow (gpm)} \times \text{Head (ft)} \times \text{Specific gravity}}{3960} \tag{14.11}$$

When calculating horsepower for positive-displacement pumps, common practice is to use psi for pressure. The formula for hydraulic horsepower then becomes

$$\text{whp} = \frac{\text{Flow (gpm)} \times \text{Pressure (psi)}}{1714} \tag{14.12}$$

Pump Efficiency and Brake Horsepower

When a motor–pump combination is used (for any purpose), neither the pump nor the motor will be 100% efficient. Simply, not all of the power supplied by the motor to the pump (*brake horsepower,* bhp) will be used to lift the water (*water* or *hydraulic horsepower,* whp); some of the power is used to overcome friction within the pump.

Similarly, not all of the power of the electric current driving the motor (*motor horsepower*, mhp) will be used to drive the pump; some of the current is used to overcome friction within the motor, and some current is lost in the conversion of electrical energy to mechanical power.

> ***Note:*** Depending on size and type, pumps are usually 50 to 85% efficient, and motors are usually 80 to 95% efficient. The efficiency of a particular motor or pump is given in the manufacturer's technical manual accompanying the unit.

The brake horsepower of a pump is equal to the hydraulic horsepower divided by the pump's efficiency. Thus, the horsepower formulas become

$$\text{bhp} = \frac{\text{Flow (gpm)} \times \text{Head (ft)} \times \text{Specific gravity}}{3960 \times \text{Efficiency}} \tag{14.13}$$

$$\text{bhp} = \frac{\text{Flow (gpm)} \times \text{Pressure (psig)}}{1714 \times \text{Efficiency}} \tag{14.14}$$

■ Example 14.9

Problem: Calculate the bhp requirements for a pump handling saltwater and having a flow of 600 gpm with 40-psi differential pressure. The specific gravity of saltwater at 68°F is 1.03. The pump efficiency is 85%.

Solution: To use Equation 14.13, convert the pressure differential to total differential head:

$$\text{Total differential head} = (40 \times 2.31) \div 1.03 = 90 \text{ ft}$$

Using Equation 14.13,

$$\text{bhp} = \frac{600 \times 90 \times 1.03}{3960 \times 0.85} = 16.5 \text{ hp}$$

Using Equation 14.14,

$$\text{bhp} = \frac{600 \times 40}{1714 \times 0.85} = 16.5 \text{ hp}$$

> ***Note:*** Horsepower requirements vary with flow. Generally, if the flow is greater, the horsepower required to move the water would be greater.

When the motor, brake, and water horsepower are known and the efficiency is unknown, a calculation to determine motor or pump efficiency must be done. Equation 14.15 is used to determine percent efficiency:

$$\text{Percent efficiency} = \frac{\text{Horsepower output}}{\text{Horsepower input}} \times 100 \tag{14.15}$$

From Equation 14.15, the specific equations to be used for motor, pump, and overall efficiency are

$$\text{Percent motor efficiency} = \frac{\text{Brake horsepower (bhp)}}{\text{Motor horsepower (mhp)}} \times 100 \qquad (14.16)$$

$$\text{Percent pump efficiency} = \frac{\text{Water horsepower (whp)}}{\text{Brake horsepower (bhp)}} \times 100 \qquad (14.17)$$

$$\text{Percent overall efficiency} = \frac{\text{Water horsepower (whp)}}{\text{Motor horsepower (mhp)}} \times 100 \qquad (14.18)$$

■ Example 14.10

Problem: A pump has a water horsepower requirement of 8.5 whp. If the motor supplies the pump with 12 hp, what is the efficiency of the pump?

Solution:

$$\text{Percent pump efficiency} = \frac{\text{Water horsepower}}{\text{Brake horsepower}} \times 100 = \frac{8.5 \text{ whp}}{12 \text{ bhp}} \times 100 = 71\%$$

■ Example 14.11

Problem: What is the efficiency if an electric power equivalent to 25 hp is supplied to the motor and 14 hp of work is accomplished by the pump?

Solution:

$$\text{Percent overall efficiency} = \frac{\text{Water horsepower}}{\text{Motor horsepower}} \times 100 = \frac{14 \text{ whp}}{25 \text{ mhp}} \times 100 = 56\%$$

■ Example 14.12

Problem: Suppose 12 kW of power are supplied to a motor. If the brake horsepower is 14 hp, what is the efficiency of the motor?

Solution: First convert the kilowatts power to horsepower. Based on the fact that 1 hp = 0.746 kW, the equation becomes

$$12 \text{ kW} \div 0.746 \text{ kW/hp} = 16.09 \text{ hp}$$

Now calculate the percent efficiency of the motor:

$$\text{Percent motor efficiency} = \frac{\text{Brake horsepower}}{\text{Motor horsepower}} \times 100 = \frac{14 \text{ bhp}}{16.09 \text{ mhp}} = 87\%$$

Specific Speed

Specific speed (N_s) refers to the speed of an impeller when pumping 1 gpm of liquid at a differential head of 1 ft. Use the following equation for specific speed, where H is at the best efficiency point:

$$N_s = \frac{\text{rpm} \times Q^{0.5}}{H^{0.75}} \tag{14.19}$$

where

rpm = Revolutions per minute.
Q = Flow (gpm).
H = Head (ft).

Pump specific speeds vary between pumps. No absolute rule sets the specific speed for different kinds of centrifugal pumps; however, the following N_s ranges are quite common:

Volute, diffuser, and vertical turbine = 500–5000
Mixed flow = 5000–10,000
Propeller pumps = 9000–15,000

Note: The higher the specific speed of a pump, the higher its efficiency.

POSITIVE-DISPLACEMENT PUMPS

The clearest differentiation between centrifugal (kinetic) pumps and positive-displacement pumps can be made based on the method by which the pumping energy is transmitted to the liquid. Centrifugal (kinetic) pumps rely on a transformation of kinetic energy to static pressure. Positive-displacement pumps, on the other hand, discharge a given volume for each stroke or revolution (that is, energy is added intermittently to the fluid flow). The two most common forms of positive-displacement pumps are reciprocating action pumps (which use pistons, plungers, diaphragms, or bellows) and rotary action pumps (using vanes, screws, lobes, or progressing cavities). No matter which form is used, all positive-displacement pumps act to force liquid into a system regardless of the resistance that may oppose the transfer. The discharge pressure generated by a positive-displacement pump is, in theory, infinite.

The three basic types of positive-displacement pumps that apply to this discussion are

- Reciprocating pumps
- Rotary pumps
- Special-purpose pumps (peristaltic or tubing pumps)

Volume of Biosolids Pumped (Capacity)

One of the most common positive-displacement biosolids pumps is the piston pump. Each stroke of a piston pump displaces or pushes out biosolids. Normally, the piston pump is operated at about 50 gpm. When making positive-displacement pump capacity calculations, we use the volume of biosolids pumped equation shown below:

$$\begin{aligned} \text{Biosolids pumped (gpm)} &= \text{Gallons pumped per stroke} \\ &\quad \times \text{No. of strokes per minute} \\ &= \left(0.785 \times D^2 \times \text{Stroke length} \times 7.48 \text{ gal/ft}^3\right) \\ &\quad \times \text{No. of strokes per minute} \end{aligned} \tag{14.20}$$

■ Example 14.13

Problem: A biosolids pump has a bore of 6 in. and a stroke length of 4 in. If the pump operates at 50 strokes (or revolutions) per minute, how many gpm are pumped? (Assume 100% efficiency.)

Solution:

$$\begin{aligned} \text{Volume} &= \left(0.785 \times D^2 \times \text{Stroke length} \times 7.48 \text{ gal/ft}^3\right) \times \text{No. of strokes per minute} \\ &= \left(0.785 \times (0.5 \text{ ft})^2 \times 0.33 \text{ ft} \times 7.48 \text{ gal/ft}^3\right) \times 50 \text{ strokes/minute} \\ &= 0.48 \text{ gal/stroke} \times 50 \text{ strokes/minute} \\ &= 25 \text{ gal/min} \end{aligned}$$

■ Example 14.14

Problem: A biosolids pump has a bore of 6 in. and a stroke setting of 3 in. The pump operates at 50 revolutions per minute. If the pump operates a total of 60 min during a 24-hr period, what is the gpd pumping rate? (Assume the piston is 100% efficient.)

Solution: First calculate the gpm pumping rate:

$$\begin{aligned} \text{Volume} &= \left(0.785 \times D^2 \times \text{Stroke length} \times 7.48 \text{ gal/ft}^3\right) \times \text{No. of strokes per minute} \\ &= \left(0.785 \times (0.5 \text{ ft})^2 \times 0.25 \text{ ft} \times 7.48 \text{ gal/ft}^3\right) \times 50 \text{ strokes/minute} \\ &= 0.37 \text{ gal/stroke} \times 50 \text{ strokes/minute} \\ &= 18.5 \text{ gal/min} \end{aligned}$$

Then convert the gpm pumping rate to a gpd pumping rate, based on total minutes pumped over 24 hr:

$$18.5 \text{ gpm} \times 60 \text{ min/day} = 1110 \text{ gpd}$$

PUMPING EXAMPLES

■ Example 14.15

Problem: A pump must pump 4800 gpm against a total head of 80 ft. What horsepower will be required to do the work?

Solution:

$$\text{whp} = \frac{\text{Flow (gpm)} \times \text{Head (ft)}}{3960} = \frac{4800 \text{ gpm} \times 80 \text{ ft}}{3960} = 97 \text{ hp}$$

■ Example 14.16

Problem: If a pump is to deliver 350 gpm of water against a total head of 60 ft, and the pump has an efficiency of 85%, what horsepower must be supplied to the pump?

Solution:

$$\text{bhp} = \frac{\text{Flow (gpm)} \times \text{Head (ft)}}{3960 \times \text{Efficiency}} = \frac{350 \text{ gpm} \times 60 \text{ ft}}{3960 \times 0.85} = 6.2 \text{ hp}$$

■ Example 14.17

Problem: The manual indicates that the output of a certain motor is 30 hp. How much horsepower must be supplied to the motor if the motor is 90% efficient?

Solution:

$$\text{mhp} = \text{bhp} \div \text{Motor efficiency} = 30 \text{ hp} \div 0.90 = 33.3 \text{ hp}$$

■ Example 14.18

Problem: The water horsepower was calculated to be 15 hp. If the motor supplies the pump with 20 hp, what must be the efficiency of the pump?

Solution:

$$\text{Percent pump efficiency} = \frac{\text{Water horsepower}}{\text{Brake horsepower}} \times 100 = \frac{15 \text{ hp}}{20 \text{ hp}} \times 100 = 75\%$$

■ Example 14.19

Problem: What is the overall efficiency if 50 hp is supplied to the motor and 28 hp of work is accomplished?

Solution:

$$\text{Percent overall efficiency} = \frac{\text{Water horsepower}}{\text{Motor horsepower}} \times 100 = \frac{28 \text{ hp}}{50 \text{ hp}} \times 100 = 56\%$$

■ **EXAMPLE 14.20**

Problem: Given that 25-kW power is supplied to a motor and the brake horsepower is 30 hp, what is the efficiency of the motor?

Solution:

$$\frac{25\text{ kW}}{0.746\text{ kW/hp}} = 33.5\text{ hp}$$

$$\text{Percent motor efficiency} = \frac{\text{Brake horsepower}}{\text{Motor horsepower}} \times 100 = \frac{30\text{ hp}}{33.5\text{ hp}} \times 100 = 89.5\%$$

■ **EXAMPLE 14.21**

Problem: A pump is discharging 1000 gpm against a head of 50 ft. The wire-to-water (overall) efficiency is 75%. If the cost of power is $0.036/kWh, what is the cost of the power consumed during a run of 100 hr?

Solution:

$$\text{mhp} = \frac{\text{bhp}}{\text{Motor efficiency}} = \frac{\text{Flow (gpm)} \times \text{Head (ft)}}{3960 \times \text{Pump efficiency} \times \text{Motor efficiency}}$$

$$= \frac{1000\text{ gpm} \times 50\text{ ft}}{3960 \times 0.75} = 16.8\text{ hp}$$

$$\text{Cost (\$/hr)} = \text{mhp} \times 0.746\text{ kW/hp} \times \text{Cost (\$/kWh)}$$

$$= 16.8\text{ hp} \times 0.746\text{ kW/hp} \times \$0.036\text{/kWh}$$

$$= \$0.45\text{/hr}$$

$$\text{Total cost} = \$0.45\text{/hr} \times 100\text{ hr} = \$45.00$$

■ **EXAMPLE 14.22**

Problem: What is the horsepower for a motor that is rated at 60 amperes and 440 volts?

Solution:

$$\text{hp} = \frac{\text{Volts} \times \text{Amperes}}{746\text{ watts/hp}} = \frac{440 \times 50\text{ amperes}}{746\text{ watts/hp}} = 29.5\text{ hp}$$

■ **EXAMPLE 14.23**

Problem: Determine the power factor for a system that used 4867 watts and pulls 11 amperes at 440 volts.

Solution:

$$\text{Power factor} = \frac{\text{Watts}}{\text{Volts} \times \text{Amperes}} = \frac{4867\text{ watts}}{440\text{ volts} \times 11\text{ amperes}} = 1$$

■ Example 14.24

Problem: If a single-phase motor pulls 12 amperes at 220 volts and has a power factor of 1.1, how many kilowatts of power does it use?

Solution:

$$\frac{\text{Volts} \times \text{Amperes} \times \text{Power factor}}{1000 \text{ watts/kW}} = \frac{220 \text{ volts} \times 12 \text{ amperes} \times 1.1}{1000 \text{ watts/kW}} = 2.9 \text{ kW}$$

■ Example 14.25

Problem: How many watts of power does a three-phase motor use if it pulls 25 amperes at 440 volts and has a power factor of 0.94?

Solution: In the following, 1.732 is the square root of 3, a value used in many three-phase calculations.

$$\text{Volts} \times \text{Amperes} \times \text{Power factor} \times 1.732 = 440 \text{ volts} \times 25 \text{ amperes} \times 0.94 \times 1.732)$$

$$= 17{,}909 \text{ watts}$$

■ Example 14.26

Problem: A pump must pump 3200 gpm against a total head of 22 ft. What horsepower (water horsepower) will be required to do the work?

Solution:

$$\text{whp} = \frac{\text{Flow (gpm)} \times \text{Head (ft)}}{3960} = \frac{3200 \text{ gpm} \times 22 \text{ ft}}{3960} = 17.8 \text{ hp}$$

■ Example 14.27

Problem: A flow of 540 gpm must be pumped against a head of 42 ft. What is the horsepower required?

Solution:

$$\text{whp} = \frac{\text{Flow (gpm)} \times \text{Head (ft)}}{3960} = \frac{540 \text{ gpm} \times 42 \text{ ft}}{3960} = 5.73 \text{ hp}$$

■ Example 14.28

Problem: A pump is pumping a total head of 73.4 ft. If 875 gpm are to be pumped, what is the water horsepower requirement?

Solution:

$$\text{whp} = \frac{\text{Flow (gpm)} \times \text{Head (ft)}}{3960} = \frac{875 \text{ gpm} \times 73.4 \text{ ft}}{3960} = 16.22 \text{ hp}$$

■ **EXAMPLE 14.29**

Problem: A pump is pumping against a total head of 48 ft. If 860 gpm are to be pumped, what is the horsepower requirement?

Solution:

$$\text{whp} = \frac{\text{Flow (gpm)} \times \text{Head (ft)}}{3960} = \frac{860 \text{ gpm} \times 48 \text{ ft}}{3960} = 10.42 \text{ hp}$$

■ **EXAMPLE 14.30**

Problem: A pump is delivering a flow 840 gpm against a total head of 33.8 ft. What is the water horsepower?

Solution:

$$\text{whp} = \frac{\text{Flow (gpm)} \times \text{Head (ft)}}{3960} = \frac{840 \text{ gpm} \times 33.8 \text{ ft}}{3960} = 7.17 \text{ hp}$$

■ **EXAMPLE 14.31**

Problem: What is the water horsepower of a pump that is producing 1555 gpm against a head of 66 ft?

Solution:

$$\text{whp} = \frac{\text{Flow (gpm)} \times \text{Head (ft)}}{3960} = \frac{1555 \text{ gpm} \times 66 \text{ ft}}{3960} = 25.9 \text{ hp}$$

■ **EXAMPLE 14.32**

Problem: If a pump delivers 380 gpm of water against a total head of 90 ft, and the pump has an efficiency of 85%, what horsepower must be supplied to the pump?

Solution:

$$\text{bhp} = \frac{\text{Flow (gpm)} \times \text{Head (ft)}}{3960 \times \text{Pump efficiency}} = \frac{380 \text{ gpm} \times 90 \text{ ft}}{3960 \times 0.85} = 10.16 \text{ hp}$$

■ **EXAMPLE 14.33**

Problem: If a pump is to deliver 460 gpm of water against a total head of 85 ft, and the pump has an efficiency of 70%, what horsepower must be supplied to the pump?

Solution:

$$\text{bhp} = \frac{\text{Flow (gpm)} \times \text{Head (ft)}}{3960 \times \text{Pump efficiency}} = \frac{460 \text{ gpm} \times 85 \text{ ft}}{3960 \times 0.70} = 14.11 \text{ hp}$$

■ Example 14.34

Problem: The motor nameplate indicates that the output of a certain motor is 33 hp. How much horsepower must be supplied to the motor if the motor is 90% efficient?

Solution:

$$\text{mhp} = \frac{\text{bhp}}{\text{Motor efficiency}} = \frac{33\text{ hp}}{0.90} = 36.67\text{ hp}$$

■ Example 14.35

Problem: The motor nameplate indicates that the output of a certain motor is 25 hp. How much horsepower must be supplied to the motor if the motor is 90% efficient?

Solution:

$$\text{mhp} = \frac{\text{bhp}}{\text{Motor efficiency}} = \frac{25\text{ hp}}{0.90} = 27.78\text{ hp}$$

■ Example 14.36

Problem: A certain pumping job will require 11 whp. If the pump is 80% efficient and the motor is 74% efficient, what motor horsepower will be required?

Solution:

$$\text{bhp} = \frac{\text{whp}}{\text{Pump efficiency}} = \frac{11\text{ whp}}{0.80} = 13.75\text{ hp}$$

$$\text{mhp} = \frac{\text{bhp}}{\text{Motor efficiency}} = \frac{13.75\text{ hp}}{0.74} = 18.58\text{ hp}$$

■ Example 14.37

Problem: A certain pumping job will required 7 whp. If the pump is 80% efficient and the motor is 90% efficient, what motor horsepower will be required?

Solution:

$$\text{bhp} = \frac{\text{whp}}{\text{Pump efficiency}} = \frac{7\text{ whp}}{0.80} = 8.75\text{ hp}$$

$$\text{mhp} = \frac{\text{bhp}}{\text{Motor efficiency}} = \frac{8.75\text{ hp}}{0.90} = 9.7\text{ hp}$$

■ Example 14.38

Problem: Based on the gallons per minute to be pumped and the total head the pump must pump against, the water horsepower requirement was calculated to be 18 whp. If the motor supplies the pump with 20 hp, what must be the efficiency of the pump?

Solution:

$$\text{Pump efficiency} = \frac{\text{whp}}{\text{bhp}} \times 100 = \frac{18\text{ hp}}{20\text{ hp}} \times 100 = 90\%$$

■ Example 14.39

Problem: What is the overall efficiency if an electric power equivalent to 30 hp is supplied to the motor and 18 hp of work is accomplished?

Solution:

$$\text{Overall efficiency} = \frac{\text{whp}}{\text{bhp}} \times 100 = \frac{18\text{ hp}}{30\text{ hp}} \times 100 = 60\%$$

■ Example 14.40

Problem: Suppose that 30-kW power is supplied to a motor. If the brake horsepower is 18 bhp, what is the efficiency of the motor?

Solution:

$$30\text{ kW} \times \frac{1\text{ hp}}{0.746\text{ kW}} = 40.2\text{ hp}$$

$$\text{Motor efficiency} = \frac{\text{bhp}}{\text{mhp}} \times 100 = \frac{18\text{ bhp}}{40.2\text{ hp}} \times 100 = 44.78\%$$

■ Example 14.41

Problem: Suppose that 12-kW power is supplied to a motor. If the brake horsepower is 10 bhp, what is the efficiency of the motor?

Solution:

$$12\text{ kW} \times \frac{1\text{ hp}}{0.746\text{ kW}} = 16.08\text{ hp}$$

$$\text{Motor efficiency} = \frac{\text{bhp}}{\text{mhp}} \times 100 = \frac{10\text{ bhp}}{16.08\text{ hp}} \times 100 = 62.2\%$$

■ Example 14.42

Problem: The motor horsepower required for a particular pumping job is 40 hp. If your power cost is \$0.09/kWh, what is the cost of operating the motor for 1 hr?

Solution:

$$\begin{aligned} \text{Cost (\$/hr)} &= \text{mhp} \times 0.746\text{ kW/hp} \times \text{Cost (\$/kWh)} \\ &= 40\text{ hp} \times 0.746\text{ kW/hp} \times \$0.09/\text{kWh} \\ &= \$2.69/\text{hr} \end{aligned}$$

■ Example 14.43

Problem: The motor horsepower required for a particular pumping job is 32 hp. If your power cost is \$0.06/kWh, what is the cost of operating the motor for 1 hr?

Solution:

$$\begin{aligned}\text{Cost (\$/hr)} &= \text{mhp} \times 0.746\ \text{kW/hp} \times \text{Cost (\$/kWh)}\\ &= 32\ \text{hp} \times 0.746\ \text{kW/hp} \times \$0.06/\text{kWh}\\ &= \$1.43/\text{hr}\end{aligned}$$

■ Example 14.44

Problem: The minimum motor horsepower requirement of a particular pumping problem is 28 mhp. If the cost of power is \$0.022/kWh, what is the cost to operate the pump for 12 hr?

Solution:

$$\begin{aligned}\text{Cost (\$/hr)} &= \text{mhp} \times 0.746\ \text{kW/hp} \times \text{Cost (\$/kWh)}\\ &= 28\ \text{hp} \times 0.746\ \text{kW/hp} \times \$0.022/\text{kWh}\\ &= \$0.46/\text{hr}\end{aligned}$$

$$\text{Total cost} = \$0.46/\text{hr} \times 12\ \text{hr} = \$5.52$$

■ Example 14.45

Problem: A pump is discharging 1200 gpm against a head of 60 ft. The wire-to-water (overall) efficiency is 70%. If the cost of power is \$0.022/kWh, what is the cost of the power consumed during a week in which the pump runs 80 hr?

Solution:

$$\text{mhp} = \frac{\text{Flow (gpm)} \times \text{Head (ft)}}{3960 \times \text{Efficiency}} = \frac{1200\ \text{gpm} \times 60\ \text{ft}}{3960 \times 0.70} = 25.97\ \text{hp}$$

$$\begin{aligned}\text{Cost (\$/hr)} &= \text{mhp} \times 0.746\ \text{kW/hp} \times \text{Cost (\$/kWh)}\\ &= 25.97\ \text{hp} \times 0.746\ \text{kW/hp} \times \$0.022/\text{kWh}\\ &= \$0.43/\text{hr}\end{aligned}$$

$$\text{Total cost} = \$0.43/\text{hr} \times 80\ \text{hr} = \$34.40$$

■ Example 14.46

Problem: Given a brake horsepower of 18 hp, a motor efficiency of 86%, and a cost of \$0.014/kWh, determine the daily power cost for operating a pump.

Solution:

$$\text{mhp} = \frac{\text{bhp}}{\text{Motor efficiency}} = \frac{18\text{ hp}}{0.86} = 20.9\text{ hp}$$

$$\begin{aligned}\text{Cost (\$/hr)} &= \text{mhp} \times 0.746\text{ kW/hp} \times \text{Cost (\$/kWh)}\\ &= 20.9\text{ hp} \times 0.746\text{ kW/hp} \times \$0.014\text{/kWh}\\ &= \$0.22\text{/hr}\end{aligned}$$

$$\text{Total cost} = \$0.22\text{/hr} \times 24\text{ hr/day} = \$5.28$$

■ Example 14.47

Problem: A pump is discharging 1300 gpm against a head of 70 ft. The wire-to-water (overall) efficiency is 66%. If the cost of power is $0.033/kWh, what is the cost of the power consumed during a week in which the pump runs 80 hr?

Solution:

$$\text{mhp} = \frac{\text{Flow (gpm)} \times \text{Head (ft)}}{3960 \times \text{Efficiency}} = \frac{1300\text{ gpm} \times 70\text{ ft}}{3960 \times 0.66} = 34.8\text{ hp}$$

$$\begin{aligned}\text{Cost (\$/hr)} &= \text{mhp} \times 0.746\text{ kW/hp} \times \text{Cost (\$/kWh)}\\ &= 34.8\text{ hp} \times 0.746\text{ kW/hp} \times \$0.033\text{/kWh}\\ &= \$0.86\text{/hr}\end{aligned}$$

$$\text{Total cost} = \$0.86\text{/hr} \times 80\text{ hr} = \$68.80$$

■ Example 14.48

Problem: What would be the horsepower on a motor that is rated at 35 amperes and 440 volts?

Solution:

$$\frac{\text{Volts} \times \text{Amperes}}{746\text{ watts/hp}} = \frac{440\text{ volts} \times 35\text{ amperes}}{746\text{ watts/hp}} = 20.64\text{ hp}$$

■ Example 14.49

Problem: What would be the horsepower on a motor that is rated at 10 amperes and 440 volts?

Solution:

$$\frac{\text{Volts} \times \text{Amperes}}{746\text{ watts/hp}} = \frac{440\text{ volts} \times 10\text{ amperes}}{746\text{ watts/hp}} = 5.9\text{ hp}$$

■ Example 14.50

Problem: What would be the horsepower on a motor that is rated at 11 amperes and 440 volts?

Solution:

$$\frac{\text{Volts} \times \text{Amperes}}{746\text{ watts/hp}} = \frac{440\text{ volts} \times 11\text{ amperes}}{746\text{ watts/hp}} = 6.49\text{ hp}$$

■ Example 14.51

Problem: How many watts of power does a single-phase motor use if it pulls 10 amperes at 110 volts and has a power factor of 1?

Solution:

$$\text{Volts} \times \text{Amperes} \times \text{Power factor} = 110\text{ volts} \times 10\text{ amperes} \times 1 = 1100\text{ watts}$$

■ Example 14.52

Problem: How many watts of power does a single-phase motor use if it pulls 10 amperes at 220 volts and has a power factor of 0.8?

Solution:

$$\text{Volts} \times \text{Amperes} \times \text{Power factor} = 220\text{ volts} \times 10\text{ amperes} \times 0.8 = 1760\text{ watts}$$

■ Example 14.53

Problem: How many watts of power does a single-phase motor use if it pulls 11 amperes at 110 volts and has a power factor of 0.5?

Solution:

$$\text{Volts} \times \text{Amperes} \times \text{Power factor} = 110\text{ volts} \times 11\text{ amperes} \times 0.5 = 605\text{ watts}$$

■ Example 14.54

Problem: How many watts of power does a three-phase motor use if it pulls 22 amperes at 440 volts and has a power factor of 0.8?

Solution:

$$\begin{aligned}\text{Volts} \times \text{Amperes} \times \text{Power factor} \times 1.732 &= 440\text{ volts} \times 22\text{ amperes} \times 0.8 \times 1.732\\ &= 13{,}412.61\text{ watts}\end{aligned}$$

■ Example 14.55

Problem: How many watts of power does a three-phase motor use if it pulls 42 amperes at 440 volts and has a power factor of 0.9?

Solution:

$$\text{Volts} \times \text{Amperes} \times \text{Power factor} \times 1.732 = 440 \text{ volts} \times 42 \text{ amperes} \times 0.9 \times 1.732$$
$$= 28{,}806.62 \text{ watts}$$

■ Example 14.56

Problem: How many kilowatts of power does a three-phase motor use if it pulls 23 amperes at 440 volts and has a power factor of 0.9?

Solution:

$$\frac{\text{Volts} \times \text{Amperes} \times \text{Power factor} \times 1.732}{1000 \text{ watts/kW}} = \frac{440 \text{ volts} \times 23 \text{ amperes} \times 0.9 \times 1.732}{1000 \text{ watts/kW}}$$
$$= 15.78 \text{ kW}$$

■ Example 14.57

Problem: What is the power factor on a system that uses 3853 watts and pulls 10 amperes at 440 volts?

Solution:

$$\text{Power factor} = \frac{\text{Watts}}{\text{Volts} \times \text{Amperes}} = \frac{3853 \text{ watts}}{440 \text{ volts} \times 10 \text{ amperes}} = 0.88$$

■ Example 14.58

Problem: What is the power factor on a system that uses 3888 watts and pulls 122 amperes at 440 volts?

Solution:

$$\text{Power factor} = \frac{\text{Watts}}{\text{Volts} \times \text{Amperes}} = \frac{3888 \text{ watts}}{440 \text{ volts} \times 12 \text{ amperes}} = 0.74$$

■ Example 14.59

Problem: During a 60-min pumping test, 9250 gal are pumped into a tank that has a length of 12 ft, width of 8 ft, and depth of 6 ft. The tank was empty before the pumping test was started. What is the gpm pumping rate?

Solution:

$$9250 \text{ gal} \div 60 \text{ min} = 154.2 \text{ gpm}$$

■ EXAMPLE 14.60

Problem: During a 30-min pumping test, 3760 gal are pumped into a tank, which has a diameter of 12 ft. The water level before the pumping test was 2 ft. What is the gpm pumping rate?

Solution:

$$3760 \text{ gal} \div 30 \text{ min} = 125.3 \text{ gpm}$$

■ EXAMPLE 14.61

Problem: A 60-ft-diameter tanks has water to a depth of 8 ft. The inlet valve is closed and a 3-hr pumping test is begun. If the water level in the tank at the end of the test is 2.5 ft, what is the gpm pumping rate?

Solution:

$$\text{Volume} = 0.785 \times (60 \text{ ft})^2 \times (8 - 2.5 \text{ ft}) = 15{,}543 \text{ ft}^3$$

$$15{,}543 \text{ ft}^3 \times 7.48 \text{ gal/ft}^3 = 116{,}261.64 \text{ gal}$$

$$\frac{116{,}261.64 \text{ gal}}{180 \text{ min}} = 645.9 \text{ gpm}$$

■ EXAMPLE 14.62

Problem: A tank has a length of 12 ft, a depth of 12 ft, and a width of 12 ft, and it has water to a depth of 8 ft. If the tank can be emptied in 1 hour, 45 minutes, what is the gpm pumping rate?

Solution:

$$\text{Volume} = (12 \text{ ft})^2 \times (8 \text{ ft}) \times 7.48 \text{ gal/ft}^3 = 8616.9 \text{ gal}$$

$$\frac{8616.9 \text{ gal}}{105 \text{ min}} = 82.07 \text{ gpm}$$

■ EXAMPLE 14.63

Problem: During a pumping test, water was pumped into an empty tank 10 ft by 10 ft by 8 ft deep. The tank completely filled with water in 12 minutes. Calculate the gpm pumping rate.

Solution:

$$\text{Volume} = (10 \text{ ft})^2 \times (8 \text{ ft}) \times 7.48 \text{ gal/ft}^3 = 5984 \text{ gal}$$

$$\frac{5984 \text{ gal}}{12 \text{ min}} = 498.7 \text{ gpm}$$

■ EXAMPLE 14.64

Problem: During a 75-minute pumping test, 12,103 gal are pumped into a tank that has a length of 16 ft, a width of 10 ft, and a depth of 8 ft. The tank was empty before the pumping test started. What is the gpm pumping rate?

Solution:

$$12{,}103 \text{ gal} \div 75 \text{ min} = 161.37 \text{ gpm}$$

15 Water Source and Storage Calculations

Approximately 40 million cubic miles of water cover or reside within the Earth. The oceans contain about 97% of all water on Earth. The other 3% is freshwater: (1) snow and ice on the surface of the Earth contain about 2.25% of the water, (2) usable groundwater is approximately 0.3%, and (3) surface freshwater is less than 0.5%. In the United States, for example, average rainfall is approximately 2.6 ft (a volume of 5900 km^3). Of this amount, approximately 71% evaporates (about 4200 km^3), and 29% goes to stream flow (about 1700 km^3).

Beneficial freshwater uses include manufacturing, food production, domestic and public needs, recreation, hydroelectric power production, and flood control. Stream flow withdrawn annually is about 7.5% (440 km^3). Irrigation and industry use almost half of this amount (3.4%, or 200 km^3/yr). Municipalities use only about 0.6% (35 km^3/yr) of this amount. Historically, in the United States, water usage has been increasing (as might be expected); for example, in 1900, 40 billion gallons of freshwater were used. In 1975, usage increased to 455 billion gallons. Estimated use in 2000 was about 720 billion gallons.

The primary sources of freshwater include the following:

- Captured and stored rainfall in cisterns and water jars
- Groundwater from springs, artesian wells, and drilled or dug wells
- Surface water from lakes, rivers, and streams
- Desalinized seawater or brackish groundwater
- Reclaimed wastewater

WATER SOURCE CALCULATIONS

Water source calculations covered in this section apply to wells and pond or lake storage capacity. Specific well calculations discussed include well drawdown, well yield, specific yield, well casing disinfection, and deep-well turbine pump capacity.

WELL DRAWDOWN

Drawdown is the drop in the level of water in a well when water is being pumped (see Figure 15.1). Drawdown is usually measured in feet or meters. One of the most important reasons for measuring drawdown is to make sure that the source water is adequate and not being depleted. The data collected to calculate drawdown can indicate if the water supply is slowly declining. Early detection can give the system

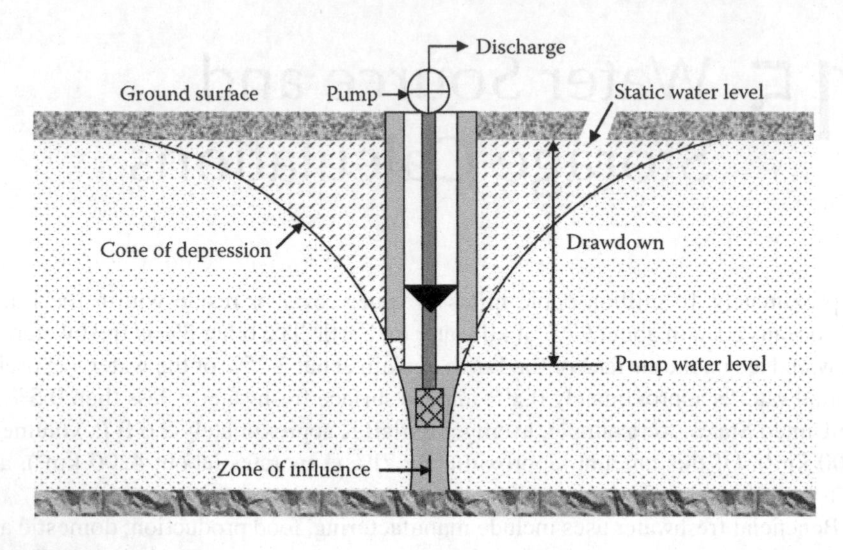

FIGURE 15.1 Drawdown.

time to explore alternative sources, establish conservation measures, or obtain any special funding that may be needed to get a new water source. Well drawdown is the difference between the pumping water level and the static water level:

$$\text{Drawdown (ft)} = \text{Pumping water level (ft)} - \text{Static water level (ft)} \quad (15.1)$$

■ Example 15.1

Problem: The static water level for a well is 70 ft. If the pumping water level is 90 ft, what is the drawdown?

Solution:

$$\text{Drawdown} = \text{Pumping water level (ft)} - \text{Static water level (ft)} = 90\text{ ft} - 70\text{ ft} = 20\text{ ft}$$

■ Example 15.2

Problem: The static water level of a well is 122 ft. The pumping water level is determined using the sounding line. The air pressure applied to the sounding line is 4.0 psi, and the length of the sounding line is 180 ft. What is the drawdown?

Solution: Calculate the water depth in the sounding line and the pumping water level:

1. Water depth in sounding line = 4.0 psi × 2.31 ft/psi = 9.2 ft
2. Pumping water level = 180 ft – 9.2 ft = 170.8 ft

Then calculate drawdown as usual:

$$\text{Drawdown} = \text{Pumping water level (ft)} - \text{Static water level (ft)}$$

$$= 170.8\text{ ft} - 122\text{ ft} = 48.8\text{ ft}$$

Well Yield

Well yield is the volume of water per unit of time that is produced from the well pumping. Usually, well yield is measured in terms of gallons per minute (gpm) or gallons per hour (gph). Sometimes, large flows are measured in cubic feet per second (cfs). Well yield is determined by using the following equation:

$$\text{Well yield (gpm)} = \text{Gallons produced} \div \text{Duration of test (min)} \tag{15.2}$$

■ Example 15.3

Problem: When the drawdown level of a well was stabilized, it was determined that the well produced 400 gal during a 5-min test. What was the well yield?

Solution:

$$\text{Well yield} = \text{Gallons produced} \div \text{Duration of test (min)} = 400\text{ gal} \div 5\text{ min} = 80\text{ gpm}$$

■ Example 15.4

Problem: During a 5-min test for well yield, a total of 780 gal was removed from the well. What was the well yield in gpm? In gph?

Solution:

$$\text{Well yield} = \text{Gallons produced} \div \text{Duration of test (min)} = 780\text{ gal} \div 5\text{ min} = 156\text{ gpm}$$

Then convert gpm flow to gph flow:

$$156\text{ gpm} \times 60\text{ min/hr} = 9360\text{ gph}$$

Specific Yield

Specific yield is the discharge capacity of the well per foot of drawdown. The specific yield may range from 1 gpm/ft drawdown to more than 100 gpm/ft drawdown for a properly developed well. Specific yield is calculated using Equation 15.3:

$$\text{Specific yield (gpm/ft)} = \text{Well yield (gpm)} \div \text{Drawdown (ft)} \tag{15.3}$$

■ Example 15.5

Problem: A well produces 260 gpm. If the drawdown for the well is 22 ft, what is the specific yield in gpm/ft?

Solution:

$$\text{Specific yield} = \text{Well yield (gpm)} \div \text{Drawdown (ft)} = 260\text{ gpm} \div 22\text{ ft} = 11.8\text{ gpm/ft}$$

■ Example 15.6

Problem: The yield for a particular well is 310 gpm. If the drawdown for this well is 30 ft, what is the specific yield in gpm/ft?

Solution:

$$\text{Specific yield} = \text{Well yield (gpm)} \div \text{Drawdown (ft)} = 310\text{ gpm} \div 30\text{ ft} = 10.3\text{ gpm/ft}$$

Well Casing Disinfection

A new, cleaned, or a repaired well normally contains contamination that may remain for weeks unless the well is thoroughly disinfected. This may be accomplished by using ordinary bleach at a concentration of 100 parts per million (ppm) of chlorine. The amount of disinfectant required is determined by the amount of water in the well. The following equation is used to calculate the pounds of chlorine required for disinfection:

$$\text{Chlorine (lb)} = \text{Chlorine (mg/L)} \times \text{Casing volume (MG)} \times 8.34\text{ lb/gal} \quad (15.4)$$

■ Example 15.7

Problem: A new well is to be disinfected with chlorine at a dosage of 50 mg/L. If the well casing diameter is 8 in. and the length of the water-filled casing is 110 ft, how many pounds of chlorine will be required?

Solution: First calculate the volume of the water-filled casing:

$$0.785 \times 0.67 \times 67 \times 110\text{ ft} \times 7.48\text{ gal/ft}^3 = 290\text{ gal}$$

Then determine the pounds of chlorine required using the mg/L to lb equation:

$$\text{Chlorine} = \text{Chlorine (mg/L)} \times \text{Volume (MG)} \times 8.34\text{ lb/gal}$$

$$= 50\text{ mg/L} \times 0.000290\text{ MG} \times 8.34\text{ lb/gal} = 0.12\text{ lb}$$

Deep-Well Turbine Pump Calculations

The deep-well turbine pump is used for high-capacity deep wells. The pump, consisting usually of more than one stage of centrifugal pump, is fastened to a pipe called the *pump column*; the pump is located in the water. The pump is driven from the surface through a shaft running inside the pump column. The water is discharged from the pump up through the pump column to the surface. The pump may be driven by a vertical shaft, electric motor at the top of the well, or some other power source, usually through a right-angle gear drive located at the top of the well. A modern version of the deep-well turbine pump is the submersible type of pump, where the pump (as well as a close-coupled electric motor built as a single unit) is located below water level in the well. The motor is built to operate submerged in water.

Vertical Turbine Pump Calculations

The calculations pertaining to well pumps include head, horsepower, and efficiency calculations. *Discharge head* is measured to the pressure gauge located close to the pump discharge flange. The pressure (psi) can be converted to feet of head using the following equation:

$$\text{Discharge head (ft)} = \text{Pressure (psi)} \times 2.31 \text{ ft/psi} \tag{15.5}$$

Total pumping head (*field head*) is a measure of the lift below the discharge head pumping water level (*discharge head*). Total pumping head is calculated as follows:

$$\text{Total pumping head (ft)} = \text{Pumping water level (ft)} + \text{Discharge head (ft)} \tag{15.6}$$

■ Example 15.8

Problem: The pressure gauge reading at a pump discharge head is 4.1 psi. What is this discharge head expressed in feet?

Solution:

$$4.1 \text{ psi} \times 2.31 \text{ ft/psi} = 9.5 \text{ ft}$$

■ Example 15.9

Problem: The static water level of a pump is 100 ft. The well drawdown is 26 ft. If the gauge reading at the pump discharge head is 3.7 psi, what is the total pumping head?

Solution:

$$\begin{aligned}\text{Total pumping head (ft)} &= \text{Pumping water level (ft)} + \text{Discharge head (ft)}\\ &= (100 \text{ ft} + 26 \text{ ft}) + (3.7 \text{ psi} \times 2.31 \text{ ft/psi})\\ &= 126 \text{ ft} + 8.5 \text{ ft} = 134.5 \text{ ft}\end{aligned}$$

There are five types of horsepower calculations for vertical turbine pumps. It is important to have a clear and retained understanding of these five types of horsepower (refer to Figure 15.2):

- *Motor horsepower* refers to the horsepower supplied to the motor. The following equation is used to calculate motor horsepower:

$$\text{Motor horsepower (input hp)} = \frac{\text{Field brake horsepower (bhp)}}{\text{Motor efficiency}/100} \tag{15.7}$$

- *Total brake horsepower* (*bhp*) refers to the horsepower output of the motor. The following equation is used to calculate total brake horsepower:

$$\text{Total bhp} = \text{Field bhp} + \text{Thrust bearing loss (hp)} \tag{15.8}$$

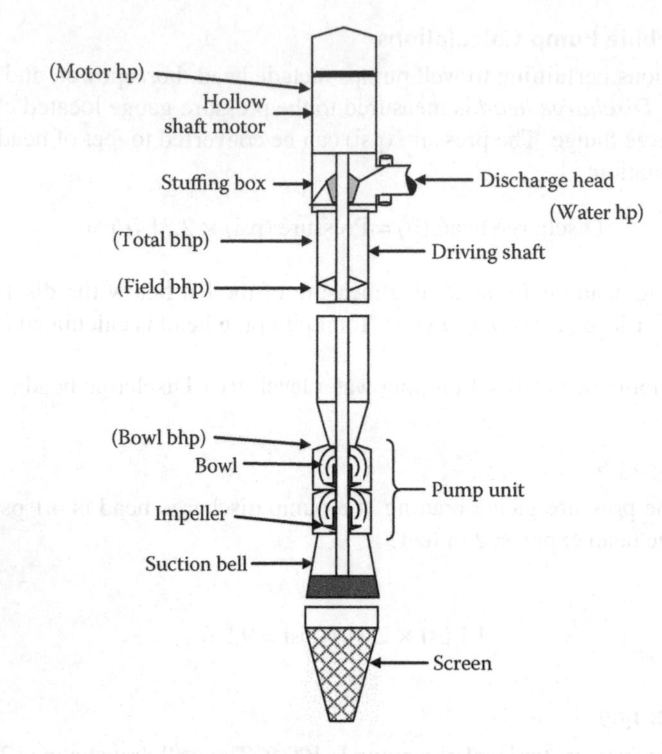

FIGURE 15.2 Vertical turbine pump, showing five horsepower types.

- *Field horsepower* refers to the horsepower required at the top of the pump shaft. The following equation is used to calculate field horsepower:

$$\text{Field bhp} = \text{Bowl bhp} + \text{Shaft loss (hp)} \tag{15.9}$$

- *Bowl* or *laboratory horsepower* refers the horsepower at the entry to the pump bowls. The following equation is used to calculate bowl horsepower:

$$\text{Bowl bhp (lab bhp)} = \frac{\text{Bowl head (ft)} \times \text{Capacity (gpm)}}{3960 \times (\text{Bowl efficiency}/100)} \tag{15.10}$$

- *Water horsepower* (*whp*) refers to the horsepower at the pump discharge. The following equation is used to calculate water horsepower:

$$\text{Water hp} = \frac{\text{Field head (ft)} \times \text{Capacity (gpm)}}{3960} \tag{15.11}$$

or the equivalent equation

$$\text{Water hp} = \frac{\text{Field head (ft)} \times \text{Capacity (gpm)}}{33{,}000 \text{ ft-lb/min}}$$

■ Example 15.10

Problem: The pumping water level for a well pump is 150 ft and the discharge pressure measured at the pump discharge centerline is 3.5 psi. If the flow rate from the pump is 700 gpm, use Equation 15.11 to calculate the water horsepower.

Solution: First calculate the field head. The discharge head must be converted from psi to ft:

$$3.5 \text{ psi} \times 2.31 \text{ ft/psi} = 8.1 \text{ ft}$$

The field head, therefore, is

$$150 \text{ ft} + 8.1 \text{ ft} = 158.1 \text{ ft}$$

The water horsepower can now be determined:

$$\text{Water hp} = \frac{\text{Field head (ft)} \times \text{Capacity (gpm)}}{33{,}000 \text{ ft-lb/min}}$$

$$= \frac{158.1 \text{ ft} \times 700 \text{ gpm} \times 8.34 \text{ lb/gal}}{33{,}000 \text{ ft-lb/min}} = 28 \text{ whp}$$

■ Example 15.11

Problem: The pumping water level for a pump is 170 ft. The discharge pressure measured at the pump discharge head is 4.2 psi. If the pump flow rate is 800 gpm, use Equation 5.11 to calculate the water horsepower.

Solution: The field head must first be determined. To determine field head, the discharge head must be converted from psi to ft:

$$4.2 \text{ psi} \times 2.31 \text{ ft/psi} = 9.7 \text{ ft}$$

The field head can now be calculated:

$$170 \text{ ft} + 9.7 \text{ ft} = 179.7 \text{ ft}$$

And the water horsepower can be calculated:

$$\text{Water hp} = \frac{179.7 \text{ ft} \times 800 \text{ gpm}}{3960} = 36 \text{ whp}$$

■ Example 15.12

Problem: A deep-well vertical turbine pump delivers 600 gpm. The lab head is 185 ft and the bowl efficiency is 84%. What is the bowl horsepower?

Solution: Use Equation 15.11 to calculate the bowl horsepower:

$$\text{Bowl bhp} = \frac{\text{Bowl head (ft)} \times \text{Capacity (gpm)}}{3960 \times (\text{Bowl efficiency}/100)}$$

$$= \frac{185\text{ ft} \times 600\text{ gpm}}{3960 \times (84/100)} = \frac{185 \times 600}{3960 \times 0.84} = 33.4\text{ bowl hp}$$

■ Example 15.13

Problem: The bowl bhp is 51.8. If the 1-in.-diameter shaft is 170 ft long and is rotating at 960 rpm with a shaft friction loss of 0.29 hp loss per 100 ft, what is the field bhp?

Solution: Before field bhp can be calculated, the shaft loss must be factored in:

$$\frac{0.29\text{ hp loss} \times 170\text{ ft}}{100} = 0.5\text{ hp loss}$$

Now the field bhp can be determined:

$$\text{Field bhp} = \text{Bowl bhp} + \text{Shaft loss (hp)} = 51.8 + 0.5 = 52.3\text{ bhp}$$

■ Example 15.14

Problem: The field horsepower for a deep-well turbine pump is 62 bhp. If the thrust bearing loss is 0.5 hp and the motor efficiency is 88%, what is the motor input horsepower?

Solution:

$$\text{Motor hp} = \frac{\text{Total bhp}}{\text{Motor efficiency}/100} = \frac{62\text{ bhp} + 0.5\text{ hp}}{0.88} = 71\text{ mhp}$$

When we speak of the *efficiency* of any machine, we are speaking primarily of a comparison of what is put out by the machine (e.g., energy output) compared to its input (e.g., energy input). Horsepower efficiency, for example, is a comparison of horsepower output of the unit or system with horsepower input to that unit or system—the unit's efficiency. There are four types efficiencies considered with vertical turbine pumps:

- Bowl efficiency
- Field efficiency
- Motor efficiency
- Overall efficiency

The general equation used in calculating percent efficiency is shown below:

$$\text{Percent (\%)} = \frac{\text{Part}}{\text{Whole}} \times 100 \tag{15.12}$$

Vertical turbine pump bowl efficiency is easily determined using a pump performance curve chart provided by the pump manufacturer.

Field efficiency is determined by

$$\text{Field efficiency (\%)} = \frac{[\text{Field head (ft)} \times \text{Capacity (gpm)}]/3960}{\text{Total bhp}} \times 100 \tag{15.13}$$

■ EXAMPLE 15.15

Problem: Given the data below, calculate the field efficiency of the deep-well turbine pump:

Field head = 180 ft
Capacity = 850 gpm
Total bhp = 61.3 bhp

Solution:

$$\text{Field efficiency (\%)} = \frac{\text{Field head (ft)} \times \text{Capacity (gpm)}}{3960 \times \text{Total bhp}} \times 100$$

$$= \frac{180\text{ ft} \times 850\text{ gpm}}{3960 \times 61.3\text{ bhp}} \times 100 = 63\%$$

The *overall efficiency* is a comparison of the horsepower output of the system with that entering the system. Equation 15.14 is used to calculate overall efficiency:

$$\text{Overall efficiency (\%)} = \frac{\text{Field efficiency (\%)} \times \text{Motor efficiency (\%)}}{100} \tag{15.14}$$

■ EXAMPLE 15.16

Problem: The efficiency of a motor is 90%. If the field efficiency is 83%, what is the overall efficiency of the unit?

Solution:

$$\text{Overall efficiency} = \frac{\text{Field efficiency (\%)} \times \text{Motor efficiency (\%)}}{100} = \frac{83 \times 90}{100} = 74.7\%$$

WATER STORAGE

Water storage facilities for water distribution systems are required primarily to provide for fluctuating demands of water usage (to provide a sufficient amount of water to average or equalize daily demands on the water supply system). In addition, other functions of water storage facilities include increasing operating convenience, leveling pumping requirements (to keep pumps from running 24 hours a day), decreasing power costs, providing water during power source or pump failure, providing large quantities of water to meet fire demands, providing surge relief (to reduce the surge associated with stopping and starting pumps), increasing detention time (to provide chlorine contact time and satisfy the desired contact time requirements), and blending water sources.

WATER STORAGE CALCULATIONS

The storage capacity, in gallons, of a reservoir, pond, or small lake can be estimated (see Figure 15.3) using Equation 15.15:

$$\text{Capacity} = \text{Avg. length (ft)} \times \text{Avg. width (ft)} \times \text{Avg. depth (ft)} \times 7.48 \text{ gal/ft}^3 \quad (15.15)$$

■ Example 15.17

Problem: A pond has an average length of 250 ft, an average width of 110 ft, and an estimated average depth of 15 ft. What is the estimated volume of the pond in gallons?

Solution:

$$\text{Volume} = \text{Avg. length (ft)} \times \text{Avg. width (ft)} \times \text{Avg. depth (ft)} \times 7.48 \text{ gal/ft}^3$$

$$= 250 \text{ ft} \times 110 \text{ ft} \times 15 \text{ ft} \times 7.48 \text{ gal/ft}^3 = 3{,}085{,}500 \text{ gal}$$

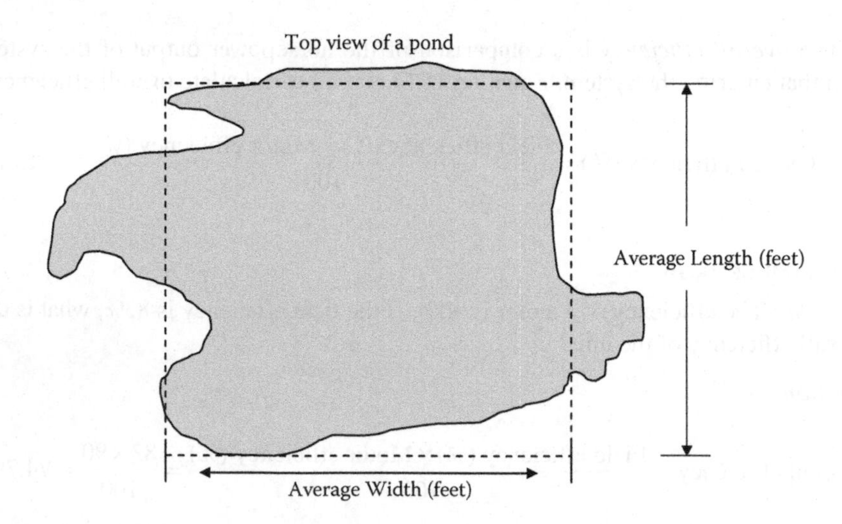

FIGURE 15.3 Determining pond storage capacity.

■ EXAMPLE 15.18

Problem: A small lake has an average length of 300 ft, an average width of 95 ft, and a maximum depth of 22 ft. What is the estimated volume of the lake in gallons?

> ***Note:*** For small ponds and lakes, the average depth is generally about 0.4 times the greatest depth; therefore, to estimate the average depth, measure the greatest depth and then multiply that number by 0.4.

Solution: First, the average depth of the lake must be estimated:

$$\text{Estimated average depth (ft)} = \text{Greatest depth (ft)} \times 0.4\text{ ft} = 22\text{ ft} \times 0.4\text{ ft} = 8.8\text{ ft}$$

Then the lake volume can be determined:

$$\text{Volume} = \text{Avg. length (ft)} \times \text{Avg. width (ft)} \times \text{Avg. depth (ft)} \times 7.48\text{ gal/ft}^3$$

$$= 300\text{ ft} \times 95\text{ ft} \times 8.8\text{ ft} \times 7.48\text{ gal/ft}^3 = 1{,}875{,}984\text{ gal}$$

COPPER SULFATE DOSING

Algal control by applying copper sulfate is perhaps the most common *in situ* treatment of lakes, ponds, and reservoirs; the copper ions in the water kill the algae. Copper sulfate application methods and dosages will vary depending on the specific surface water body being treated. The desired copper sulfate dosage may be expressed in mg/L copper, lb/ac-ft copper sulfate, or lb/ac copper sulfate.

For a dose expressed as mg/L copper, the following equation is used to calculate the pounds of copper sulfate required:

$$\text{Copper sulfate (lb)} = \frac{\text{Copper (mg/L)} \times \text{Volume (MG)} \times 8.34\text{ lb/gal}}{\text{\% Available copper/100}} \tag{15.16}$$

■ EXAMPLE 15.19

Problem: For algae control in a small pond, a dosage of 0.5 mg/L copper is desired. The pond has a volume of 15 MG. How many pounds of copper sulfate will be required? (Copper sulfate contains 25% available copper.)

Solution:

$$\text{Copper sulfate} = \frac{\text{Copper (mg/L)} \times \text{Volume (MG)} \times 8.34\text{ lb/gal}}{\text{\% Available copper/100}}$$

$$= \frac{0.5\text{ mg/L} \times 15\text{ MG} \times 8.34\text{ lb/gal}}{25/100} = 250\text{ lb}$$

For calculating lb/ac-ft copper sulfate, use the following equation to determine the pounds of copper sulfate required (assume the desired copper sulfate dosage is 0.9 lb/ac-ft):

$$\text{Copper sulfate (lb)} = \frac{\text{0.9 lb copper sulfate} \times \text{ac-ft}}{\text{1 ac-ft}} \tag{15.17}$$

■ Example 15.20

Problem: A pond has a volume of 35 ac-ft. If the desired copper sulfate dose is 0.9 lb/ac-ft, how many pounds of copper sulfate will be required?

Solution:

$$\text{Copper sulfate (lb)} = \frac{\text{0.9 lb copper sulfate} \times \text{ac-ft}}{\text{1 ac-ft}}$$

$$\frac{\text{0.9 lb copper sulfate}}{\text{1 ac-ft}} = \frac{x \text{ lb copper sulfate}}{\text{35 ac-ft}}$$

Then solve for x:

$$0.9 \times 35 = x \text{ lb}$$

$$x = 31.5 \text{ lb}$$

The desired copper sulfate dosage may also be expressed in terms of lb/ac copper sulfate. The following equation is used to determine the pounds of copper sulfate required (assume that the desired copper sulfate dosage is 5.2 lb/ac):

$$\text{Copper sulfate (lb)} = \frac{\text{5.2 lb copper sulfate} \times \text{ac}}{\text{1 ac}} \tag{15.18}$$

■ Example 15.21

Problem: A small lake has a surface area of 6.0 ac. If the desired copper sulfate dose is 5.2 lb/ac, how many pounds of copper sulfate are required?

Solution:

$$\text{Copper sulfate} = \frac{\text{5.2 lb copper sulfate} \times \text{6 ac}}{\text{1 ac}} = 31.2 \text{ lb}$$

16 Water/Wastewater Laboratory Calculations

Waterworks and wastewater treatment plants are sized to meet current needs, as well as those of the future. No matter the size of the treatment plant, some space or area within the plant is designated as the lab area, which can range from being the size of a closet to being fully equipped and staffed environmental laboratories. Water and wastewater laboratories usually perform a number of different tests. Lab test results provide the operator with the information necessary to operate the treatment facility at optimal levels. Laboratory testing usually includes determining service line flushing time, solution concentration, pH, chemical oxygen demand (COD), total phosphorus, fecal coliform count, chlorine residual, and biochemical oxygen demand (BOD), to name a few. The standard reference for performing wastewater testing is contained in *Standard Methods for the Examination of Water & Wastewater.*

In this chapter, the focus is on standard water/wastewater lab tests that involve various calculations. Specifically, the focus is on calculations used to determine the proportioning factor for composite sampling, an estimation of flow from a faucet, service line flushing time, solution concentration, BOD, molarity and moles, normality, settleability, settleable solids, total solids, fixed and volatile solids, suspended solids, volatile suspended solids, biosolids volume index, and biosolids density index.

FAUCET FLOW ESTIMATION

On occasion, the waterworks sampler must take water samples from a customer's residence. In small water systems, the sample is usually taken from the customer's front yard faucet. A convenient flow rate for taking water samples is about 0.5 gallons per minute (gpm). To estimate the flow from a faucet, use a 1-gallon container and record the time it takes to fill the container. To calculate the flow in gallons per minute, insert the recorded information into Equation 16.1:

$$\text{Flow (gpm)} = \text{Volume (gal)} \div \text{Time (min)} \quad (16.1)$$

■ Example 16.1

Problem: The flow from a faucet filled up the gallon container in 48 sec. What was the flow rate from the faucet in gallons per minute? Because the flow rate is desired in minutes the time should also be expressed as minutes:

$$48 \text{ sec} \div 60 \text{ sec/min} = 0.80 \text{ min}$$

Solution: Calculate the flow rate from the faucet:

$$\text{Flow} = \text{Volume (gal)} \div \text{Time (min)} = 1 \text{ gal} \div 0.80 \text{ min} = 1.25 \text{ gpm}$$

■ EXAMPLE 16.2

Problem: The flow from a faucet filled up a gallon container in 55 sec. What was the flow rate from the faucet in gallons per minute?

Solution:

$$55 \text{ sec} \div 60 \text{ sec/min} = 0.92 \text{ minute}$$

Calculate the flow rate:

$$\text{Flow} = \text{Volume (gal)} \div \text{Time (min)} = 1 \text{ gal} \div 0.92 \text{ min} = 1.1 \text{ gpm}$$

SERVICE LINE FLUSHING TIME

To determine the quality of potable water delivered to the consumer, a sample is taken from the customer's outside faucet—water that is typical of the water delivered. To obtain an accurate indication of the system water quality, this sample must be representative. Further, to ensure that the sample taken is typical of water delivered, the service line must be flushed twice. Equation 16.2 is used to calculate flushing time:

$$\text{Flushing time (min)} = \frac{0.785 \times D^2 \times \text{Length (ft)} \times 7.48 \text{ gal/ft}^3 \times 2}{\text{Flow rate (gpm)}} \quad (16.2)$$

■ EXAMPLE 16.3

Problem: How long (in minutes) will it take to flush a 40-ft length of a 1/2-in.-diameter service line if the flow through the line is 0.5 gpm?

Solution: Calculate the diameter of the pump in feet:

$$0.50 \text{ in.} \div 12 \text{ in./ft} = 0.04 \text{ ft}$$

Calculate the flushing time:

$$\text{Flushing time} = \frac{0.785 \times D^2 \times \text{Length (ft)} \times 7.48 \text{ gal/ft}^3 \times 2}{\text{Flow rate (gpm)}}$$

$$= \frac{0.785 \times (0.04)^2 \times 40 \text{ ft} \times 7.48 \text{ gal/ft}^3 \times 2}{0.5 \text{ gpm}}$$

$$= 1.5 \text{ min}$$

■ **EXAMPLE 16.4**

Problem: At a flow rate of 0.5 gpm, how long (in minutes and seconds) will it take to flush a 60-ft length of 3/4-in. service line?

Solution:

$$3/4\text{-inch diameter} = 0.06 \text{ ft.}$$

$$\text{Flushing time} = \frac{0.785 \times D^2 \times \text{Length (ft)} \times 7.48 \text{ gal/ft}^3 \times 2}{\text{Flow rate (gpm)}}$$

$$= \frac{0.785 \times (0.06)^2 \times 60 \text{ ft} \times 7.48 \text{ gal/ft}^3 \times 2}{0.5 \text{ gpm}}$$

$$= 5.1 \text{ min}$$

Convert the fractional part of a minute (0.1) to seconds:

$$0.1 \text{ min} \times 60 \text{ sec/min} = 6 \text{ sec}$$

$$5.1 \text{ min} = 5 \text{ min, } 6 \text{ sec}$$

COMPOSITE SAMPLING CALCULATION (PROPORTIONING FACTOR)

When preparing oven-baked food, the cook sets the correct oven temperature and then usually moves on to some other chore. The oven thermostat maintains the correct temperature, and that is that. Unlike the cook, in water and wastewater treatment plant operations the operator does not have the luxury of setting a plant parameter and then walking off and forgetting about it. To optimize plant operations, various adjustments to unit processes must be made on an ongoing basis.

The operator makes unit process adjustments based on local knowledge (experience) and on lab test results. However, before lab tests can be performed, samples must be taken. The two basic types of samples are *grab samples* and *composite samples*. The type of sample taken depends on the specific test, the reason the sample is being collected, and the requirements in the plant discharge permit.

A grab sample is a discrete sample collected at one time and one location. It is primarily used for any parameter whose concentration can change quickly (e.g., dissolved oxygen, pH, temperature, total chlorine residual) and is representative only of the conditions at the time of collection.

A composite sample consists of a series of individual grab samples taken at specified time intervals and in proportion to flow. The individual grab samples are mixed together in proportion to the flow rate at the time the sample was collected to form the composite sample. The composite sample represents the character of the water/wastewater over a period of time.

Composite Sampling Procedure and Calculation

Because knowledge of the procedure used in processing composite samples is important to water/wastewater operators, the actual procedure used is covered in this section:

1. Determine the total amount of sample required for all tests to be performed on the composite sample.
2. Determine the average daily flow of the treatment system.

Note: Average daily flow can be determined by using several months of data, which will provide a more representative value.

3. Calculate a proportioning factor:

$$\text{Proportioning factor} = \frac{\text{Total sample volume required (mL)}}{\text{No. of samples} \times \text{Average daily flow (MGD)}} \tag{16.3}$$

Note: Round the proportioning factor to the nearest 50 units (e.g., 50, 100, 150) to simplify calculation of the sample volume.

4. Collect the individual samples in accordance with the schedule (e.g., once/hr, once/15 min).
5. Determine the flow rate at the time the sample was collected.
6. Calculate the specific amount to add to the composite container:

$$\text{Required volume (mL)} = \text{Flow}_T \times \text{Proportioning factor} \tag{16.4}$$

   where T = time sample was collected.
7. Mix the individual sample thoroughly, measure the required volume, and add to a composite storage container.
8. Keep the composite sample refrigerated throughout the collection period.

■ Example 16.5

Problem: The effluent testing will require 3825 mL of sample. The average daily flow is 4.25 MGD. Using the flows given below, calculate the amount of sample to be added at each of the times shown:

8 a.m.	3.88 MGD
9 a.m.	4.10 MGD
10 a.m.	5.05 MGD
11 a.m.	5.25 MGD
12 noon	3.80 MGD
1 p.m.	3.65 MGD
2 p.m.	3.20 MGD
3 p.m.	3.45 MGD
4 p.m.	4.10 MGD

Solution:

Proportioning factor = 3825 mL ÷ (9 samples × 4.25 MGD) = 100

$Volume_{8a.m.}$ = 3.88 × 100 = 388 mL
$Volume_{9a.m.}$ = 4.10 × 100 = 410 mL
$Volume_{10a.m.}$ = 5.05 × 100 = 505 mL
$Volume_{11a.m.}$ = 5.25 × 100 = 525 mL
$Volume_{12noon}$ = 3.80 × 100 = 380 mL
$Volume_{1p.m.}$ = 3.65 × 100 = 365 mL
$Volume_{2p.m.}$ = 3.20 × 100 = 320 mL
$Volume_{3p.m.}$ = 3.45 × 100 = 345 mL
$Volume_{4p.m.}$ = 4.10 × 100 = 410 mL

BIOCHEMICAL OXYGEN DEMAND CALCULATIONS

Biochemical oxygen demand (BOD_5) measures the amount of organic matter that can be biologically oxidized under controlled conditions (5 days at 20°C in the dark). Several criteria determine which BOD_5 dilutions should be used for calculating test results. Consult a laboratory testing reference manual (such as *Standard Methods*) for this information. Two basic calculations are used for BOD_5. The first is used for samples that have not been seeded, and the second must be used whenever BOD_5 samples are seeded. Both methods are introduced and examples are provided below.

BOD_5 (Unseeded)

$$\text{BOD}_5 \text{ (unseeded)} = \frac{[\text{DO}_{\text{start}} \text{ (mg/L)} - \text{DO}_{\text{final}} \text{ (mg/L)}] \times 300 \text{ mL}}{\text{Sample volume (mL)}} \quad (16.5)$$

■ Example 16.6

Problem: A BOD_5 test has been completed. Bottle 1 of the test had dissolved oxygen (DO) of 7.1 mg/L at the start of the test. After 5 days, bottle 1 had a DO of 2.9 mg/L. Bottle 1 contained 120 mg/L of sample. Determine the unseeded BOD_5.

Solution:

$$\text{BOD}_5 \text{ (unseeded)} = \frac{(7.1 \text{ mg/L} - 2.9 \text{ mg/L}) \times 300 \text{ mL}}{120 \text{ mL}} = 10.5 \text{ mg/L}$$

BOD_5 (Seeded)

If the BOD_5 sample has been exposed to conditions that could reduce the number of healthy, active organisms, the sample must be seeded with organisms. Seeding requires the use of a correction factor to remove the BOD_5 contribution of the seed material:

$$\text{Seed correction} = \frac{\text{Seed material BOD}_5 \times \text{Seed in dilution (mL)}}{300\text{ mL}} \quad (16.6)$$

$$\text{BOD}_5\text{ (seeded)} = \frac{\left[\text{DO}_{\text{start}}\text{ (mg/L)} - \text{DO}_{\text{final}}\text{ (mg/L)} - \text{Seed correction}\right] \times 300\text{ mL}}{\text{Sample volume (mL)}} \quad (16.7)$$

■ Example 16.7

Problem: Using the data provided below, determine the BOD_5:

BOD_5 of seed material = 90 mg/L
Seed material = 3 mL
Sample = 100 mL
Start DO = 7.6 mg/L
Final DO = 2.7 mg/L

Solution:

$$\text{Seed correction} = \frac{90\text{ mg/L} \times 3\text{ mL}}{300\text{ mL}} = 0.90\text{ mg/L}$$

$$\text{BOD}_5\text{ (seeded)} = \frac{[(7.6\text{ mg/L} - 2.7\text{ mg/L}) - 0.90] \times 300\text{ mL}}{100\text{ mL}} = 12\text{ mg/L}$$

BOD 7-Day Moving Average

Because the BOD characteristic of wastewater varies from day to day, even hour to hour, operational control of the treatment system is most often accomplished based on trends in data rather than individual data points. The BOD 7-day moving average is a calculation of the BOD trend.

Note: The 7-day moving average is called a moving average because a new average is calculated each day by adding the new day's value to the six previous days' values.

$$\text{7-Day average BOD} = \frac{\text{BOD}_{\text{day1}} + \text{BOD}_{\text{day2}} + \text{BOD}_{\text{day3}} + \text{BOD}_{\text{day4}} + \text{BOD}_{\text{day5}} + \text{BOD}_{\text{day6}} + \text{BOD}_{\text{day7}}}{7} \quad (16.8)$$

■ Example 16.8

Problem: Given the following primary effluent BOD test results, calculate the 7-day average:

June 1	200 mg/L
June 2	210 mg/L
June 3	204 mg/L

June 4	205 mg/L
June 5	222 mg/L
June 6	214 mg/L
June 7	218 mg/L

Solution:

$$\text{7-day average BOD} = (200 + 210 + 204 + 205 + 222 + 214 + 218) \div 7 = 210 \text{ mg/L}$$

MOLES AND MOLARITY

Chemists have defined a very useful unit called the *mole*. Moles and molarity, a concentration term based on the mole, have many important applications in water/wastewater operations. A mole is defined as a gram molecular weight—that is, the molecular weight expressed as grams. For example, a mole of water is 18 g of water, and a mole of glucose is 180 g of glucose. A mole of any compound always contains the same number of molecules. The number of molecules in a mole is called *Avogadro's number* and has a value of 6.022×10^{23}.

> ***Note:*** How big is Avogadro's number? An Avogadro's number of soft drink cans would cover the surface of the Earth to a depth of over 200 miles.
>
> ***Note:*** Molecular weight is the weight of one molecule. It is calculated by adding the weights of all of the atoms that are present in one molecule. The units are atomic mass units (amu). A mole is a gram molecular weight—that is, the molecular weight expressed in grams. The molecular weight is the weight of one molecule in daltons. All moles contain the same number of molecules (Avogadro's number), equal to 6.022×10^{23}. The reason why all moles have the same number of molecules is because the value of the mole is proportional to the molecular weight.

MOLES

A mole is a quantity of a compound equal in weight to its formula weight; for example, the formula weight for water (see Figure 16.1) can be determined using the Periodic Table of Elements:

$$\begin{aligned} \text{Hydrogen } (1.008) \times 2 &= 2.016 \\ + \text{Oxygen} &= \underline{16.000} \\ \text{Formula weight of } H_2O &= 18.016 \end{aligned}$$

Because the formula weight of water is 18.016, a mole is 18.016 units of weight. A *gram-mole* is 18.016 grams of water. A *pound-mole* is 18.016 pounds of water. For our purposes in this text, the term *mole* will be understood to mean gram-mole. The equation used to determine moles is shown below:

$$\text{Moles} = \text{Grams of chemical} \div \text{Formula weight of chemical} \quad (16.9)$$

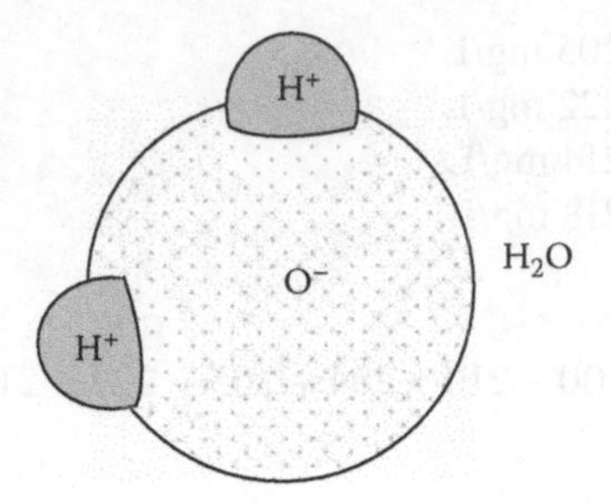

FIGURE 16.1 A molecule of water.

■ **EXAMPLE 16.9**

Problem: The atomic weight of a certain chemical is 66. If 35 g of the chemical are used in making up a 1-L solution, how many moles are used?

Solution:

$$\text{Grams of chemical} \div \text{Formula weight of chemical} = 66\text{ g} \div 35\text{ g/mole} = 1.9\text{ moles}$$

The molarity of a solution is calculated by taking the moles of solute and dividing by the liters of solution:

$$\text{Molarity} = \text{Moles of solute} \div \text{Liters of solution} \tag{16.10}$$

■ **EXAMPLE 16.10**

Problem: What is the molarity of 2 moles of solute dissolved in 1 L of solvent?

Solution:

$$\text{Molarity} = 2\text{ moles} \div 1\text{ L} = 2\,M$$

> **Note:** Measurement in *moles* is a measurement of the amount of a substance. Measurement in *molarity* is a measurement of the concentration of a substance—the amount (moles) per unit volume (liters).

NORMALITY

The molarity of a solution refers to its concentration (the solute dissolved in the solution). The normality of a solution refers to the number of *equivalents* of solute per liter of solution. The definition of chemical equivalent depends on the substance or type of chemical reaction under consideration. Because the concept of equivalents is based on the reacting power of an element or compound, it follows that a specific number of equivalents of one substance will react with the same number of equivalents of another substance. When the concept of equivalents is taken into consideration, it is less likely that chemicals will be wasted as excess amounts. Keeping in mind that normality is a measure of the reacting power of a solution (i.e., 1 equivalent of a substance reacts with 1 equivalent of another substance), we use the following equation to determine normality:

$$\text{Normality } (N) = \text{No. of equivalents of solute} \div \text{Liters of solution} \quad (16.11)$$

■ EXAMPLE 16.11

Problem: If 2.0 equivalents of a chemical are dissolved in 1.5 L of solution, what is the normality of the solution?

Solution:

$$\text{No. of equivalents of solute} \div \text{Liters of solution} = 2.0 \text{ equivalents} \div 1.5 \text{ L} = 1.33\ N$$

■ EXAMPLE 16.12

Problem: An 800-mL solution contains 1.6 equivalents of a chemical. What is the normality of the solution?

Solution: First convert 800 mL to liters:

$$800 \text{ mL} \div 1000 \text{ mL} = 0.8 \text{ L}$$

Then calculate the normality of the solution:

$$\text{No. of equivalents of solute} \div \text{Liters of solution} = 1.6 \text{ equivalents} \div 0.8 \text{ L} = 2\ N$$

SETTLEABILITY (ACTIVATED BIOSOLIDS SOLIDS)

The settleability test is a test of the quality of the activated biosolids solids—or activated sludge solids (mixed-liquor suspended solids, MLSS). Settled biosolids volume (SBV)—or settled sludge volume (SSV)—is determined at specified times during sample testing. For control, 30- and 60-minute observations are made. Subscripts (SBV_{30} or SSV_{30} and SBV_{60} or SSV_{60}) indicate settling time. A sample of activated biosolids is taken from the aeration tank, poured into a 2000-mL graduated cylinder, and allowed to settle for 30 or 60 min. The settling characteristics of the biosolids in the graduated cylinder give a general indication of the settling of the MLSS in the final clarifier. From the settleability test, the percent settleable solids can be calculated using the following equation:

$$\text{\% Settleable solids} = \frac{\text{Settled solids (mL)}}{\text{2000-mL sample}} \times 100 \quad (16.12)$$

■ **Example 16.13**

Problem: A settleability test is conducted on a sample of MLSS. What is percent settleable solids if 420 mL settle in the 2000-mL graduate?

Solution:

$$\%\ \text{Settleable solids} = \frac{\text{Settled solids (mL)}}{\text{2000-mL sample}} \times 100 = \frac{420\ \text{mL}}{2000\ \text{mL}} \times 100 = 21\%$$

■ **Example 16.14**

Problem: A 2000-mL sample of activated biosolids is tested for settleability. If the settled solids are measured as 410 mL, what is the percent settled solids?

Solution:

$$\%\ \text{Settleable solids} = \frac{\text{Settled solids (mL)}}{\text{2000-mL sample}} \times 100 = \frac{410\ \text{mL}}{2000\ \text{mL}} \times 100 = 20.5\%$$

SETTLEABLE SOLIDS

The settleable solids test is an easy, quantitative method to measure sediment found in wastewater. An Imhoff cone (see Figure 16.2) is filled with 1 L of sample wastewater, stirred, and allowed to settle for 60 min. The settleable solids test, unlike the settleability test, is conducted on samples from the sedimentation tank or clarifier influent and effluent to determine percent removal of settleable solids. The percent settleable solids is determined by the following equation:

$$\%\ \text{Settleable solids removed} = \frac{\text{Settled solids removed (mL/L)}}{\text{Settled solids in influent (mL/L)}} \times 100 \qquad (16.13)$$

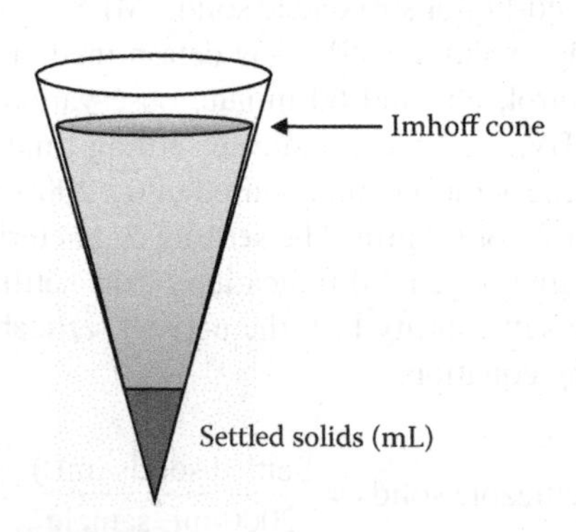

FIGURE 16.2 1-liter Imhoff cone.

■ **EXAMPLE 16.15**

Problem: Calculate the percent removal of settleable solids if the settleable solids of the sedimentation tank influent is 15 mL/L and the settleable solids of the effluent is 0.4 mL/L.

Solution: First determine the settleable solids removed:

$$15.0 \text{ mL/L} - 0.4 \text{ mL/L} = 14.6 \text{ mL/L}$$

Next, insert the parameters into Equation 16.13:

$$\% \text{ Settleable solids removed} = \frac{\text{Settled solids removed (mL/L)}}{\text{Settled solids in influent (mL/L)}} \times 100$$

$$= \frac{14.6 \text{ mL/L}}{15.0 \text{ mL/L}} \times 100 = 97\%$$

■ **EXAMPLE 16.16**

Problem: Calculate the percent removal of settleable solids if the settleable solids of the sedimentation tank influent are 13 mL/L and the settleable solids of the effluent are 0.5 mL/L.

Solution: First determine removed settleable solids:

$$13 \text{ mL/L} - 0.5 \text{ mL/L} = 12.5 \text{ mL/L}$$

Next, insert the parameters into Equation 16.13:

$$\% \text{ Settleable solids removed} = \frac{\text{Settled solids removed (mL/L)}}{\text{Settled solids in influent (mL/L)}} \times 100$$

$$= \frac{12.5 \text{ mL/L}}{13.0 \text{ mL/L}} \times 100 = 96\%$$

TOTAL SOLIDS, FIXED SOLIDS, AND VOLATILE SOLIDS

Wastewater consists of both water and solids. The *total solids* may be further classified as either *volatile solids* (organics) or *fixed solids* (inorganics) (see Figure 16.3). Normally, total solids and volatile solids are expressed as percents, whereas suspended solids are generally expressed as mg/L. To calculate either percents or mg/L concentrations, certain concepts must be understood:

- *Total solids*—The residue left in the vessel after evaporation of liquid from a sample and subsequent drying in an oven at 103 to 105°C
- *Fixed solids*—The residue left in the vessel after a sample is ignited (heated to dryness at 550°C)
- *Volatile solids*—The weight loss after a sample is ignited (heated to dryness at 550°C)

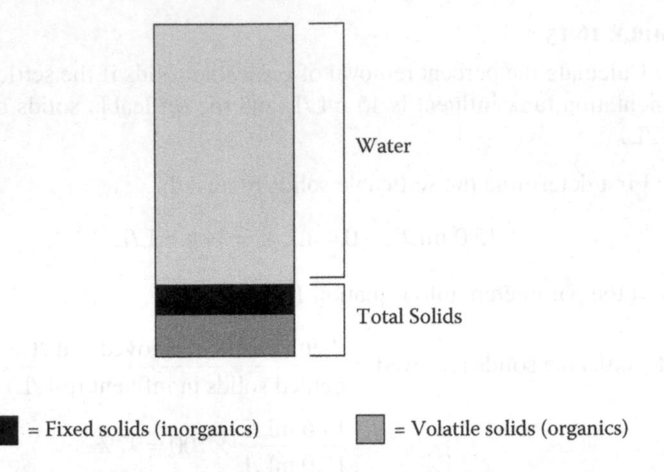

FIGURE 16.3 Composition of wastewater.

Note: When the term *biosolids* is used, it may be understood to mean a semiliquid mass composed of solids and water. The term *solids,* however, is used to mean dry solids after the evaporation of water.

The percent total solids and percent volatile solids are calculated as follows:

$$\%\ \text{Total solids} = \frac{\text{Total solids weight}}{\text{Biosolids sample weight}} \times 100 \tag{16.14}$$

$$\%\ \text{Volatile solids} = \frac{\text{Volatile solids weight}}{\text{Total solids weight}} \times 100 \tag{16.15}$$

■ Example 16.17

Problem: Given the information below, determine the percent solids in the sample and the percent of volatile solids in the biosolids sample.

	Biosolids Sample	After Drying	After Burning (Ash)
Weight of sample and dish	73.43 g	24.88	22.98
Weight of dish (tare weight)	22.28 g	22.28	22.28

Solution: To calculate the percent total solids, the grams total solids (solids after drying) and grams biosolids sample must be determined:

Total solids

24.88 g (weight of total solids and dish)
–22.28 g (weight of dish)
2.60 g (weight of total solids)

Biosolids sample

73.43 g (weight of biosolids and dish)
–22.28 g (weight of dish)
51.15 g (weight of biosolids sample)

$$\% \text{ Total solids} = \frac{\text{Total solids weight}}{\text{Biosolids sample weight}} \times 100 = \frac{2.60 \text{ g}}{51.15 \text{ g}} \times 100 = 5\%$$

To calculate the percent volatile solids, the grams total solids and grams volatile solids must be determined. Because the total solids value has already been calculated (above), only volatile solids must be calculated:

Volatile solids

24.88 g (weight of sample and dish *before* burning)
–22.98 g (weight of sample and dish *after* burning)
1.90 g (weight of solids lost in burning)

$$\% \text{ Volatile solids} = \frac{\text{Volatile solids weight}}{\text{Total solids weight}} \times 100 = \frac{1.90 \text{ g}}{2.60 \text{ g}} \times 100 = 73\%$$

WASTEWATER SUSPENDED SOLIDS AND VOLATILE SUSPENDED SOLIDS

Total suspended solids (TSS) are the amount of filterable solids in a wastewater sample. Samples are filtered through a glass fiber filter. The filters are dried and weighed to determine the amount of total suspended solids in mg/L of sample. *Volatile suspended solids* (VSS) are those solids lost on ignition (heating to 500°C). They are useful to the treatment plant operator because they give a rough approximation of the amount of organic matter present in the solid fraction of wastewater, activated biosolids, and industrial wastes. With the exception of the required drying time, the suspended solids and volatile suspended solids tests of wastewater are similar to those of the total and volatile solids performed for biosolids. Calculations of suspended solids and volatile suspended solids are demonstrated in Example 16.18.

Note: The total and volatile solids of biosolids are generally expressed as percents, by weight. The biosolids samples are 100 mL and are unfiltered.

■ EXAMPLE 16.18

Problem: Given the following information regarding a primary effluent sample, calculate the mg/L suspended solids and the percent volatile suspended solids of the 50-mL sample.

	After Drying (Before Burning)	After Burning (Ash)
Weight of sample and dish	24.6268 g	24.6232 g
Weight of dish (tare weight)	24.6222 g	24.6222 g

Solution: To calculate the milligrams suspended solids per liter of sample (mg/L), we must first determine grams suspended solids:

24.6268 g (weight of dish and suspended solids)
–24.6222 g (weight of dish)
0.0046 g (weight of suspended solids)

Next, we calculate mg/L suspended solids (using a multiplication factor of 20, a number that will vary with sample volume) to make the denominator equal to 1 L (1000 mL):

$$\text{Suspended solids} = \frac{0.0046\text{ g SS}}{50\text{ mL}} \times \frac{1000\text{ mg}}{1\text{ g}} \times \frac{20}{20} = \frac{92\text{ mg}}{1000\text{ mL}} = 92\text{ mg/L}$$

To calculate percent volatile suspended solids, we must know the weight of both total suspended solids (calculated above) and volatile suspended solids.

24.6268 g (weight of dish and suspended solids *before* burning)
–24.6234 g (weight of dish and suspended solids *after* burning)
0.0034 g (weight of solids lost in burning)

$$\%\text{ Volatile suspended solids} = \frac{\text{Volatile solids weight}}{\text{Suspended solids weight}} \times 100 = \frac{0.0034\text{ g}}{0.0046\text{ g}} \times 100 = 70\%$$

BIOSOLIDS VOLUME INDEX AND BIOSOLIDS DENSITY INDEX

Two variables are used to measure the settling characteristics of activated biosolids and to determine what the return biosolids pumping rate should be. These are the *biosolids volume index* (BVI) and the *biosolids density index* (BDI):

$$\text{BVI} = \frac{\%\text{ MLSS volume after 30 min}}{\%\text{ MLSS}} = \text{Settled biosolids (mL)} \times 1000 \quad (16.16)$$

$$\text{BDI} = \frac{\%\text{ MLSS}}{\%\text{ MLSS volume after 30 min settling}} \times 100 \quad (16.17)$$

These indices relate the weight of biosolids to the volume the biosolids occupies. They show how well the liquid–solids separation part of the activated biosolids system is performing its function on the biological floc that has been produced and is to be settled out and returned to the aeration tanks or wasted. The better the liquid–solids separation is, the smaller will be the volume occupied by the settled biosolids and the lower the pumping rate required to keep the solids in circulation.

■ **Example 16.19**

Problem: The settleability test indicates that, after 30 min, 220 mL of biosolids settle in the 1-L graduated cylinder. If the mixed-liquor suspended solids (MLSS) concentration in the aeration tank is 2400 mg/L, what is the biosolids volume index?

Solution:

$$\text{BVI} = \frac{\text{Volume}}{\text{Density}} = \frac{220 \text{ mL/L}}{2400 \text{ mg/L}} = \frac{220 \text{ mL}}{2400 \text{ mg}} = \frac{220 \text{ mL}}{2.4 \text{ g}} = 92$$

The biosolids density index is also a method of measuring the settling quality of activated biosolids, yet it, like the BVI parameter, may or may not provide a true picture of the quality of the biosolids in question unless compared with other relevant process parameters. It differs from the BVI in that the higher the BDI value, the better the settling quality of the aerated mixed liquor. Similarly, the lower the BDI, the poorer the settling quality of the mixed liquor. The BDI is the concentration in percent solids that the activated biosolids will assume after settling for 30 minutes. The BDI will range from 2.00 to 1.33, and biosolids with values of 1 or more are generally considered to have good settling characteristics. To calculate the BDI, we simply invert the numerator and denominator and multiply by 100.

■ **Example 16.20**

Problem: The MLSS concentration in the aeration tank is 2500 mg/L. If the activated biosolids settleability test indicates 225 mL settled in the 1-L graduated cylinder, what is the biosolids density index?

Solution:

$$\text{BDI} = \frac{\text{Density (determined by MLSS concentration)}}{\text{Volume (determined by settleability test)}} \times 100$$

$$= \frac{2500 \text{ mg}}{225 \text{ mL}} \times 100 = \frac{2.5 \text{ g}}{225 \text{ mL}} \times 100 = 1.11$$

CHEMICAL DOSAGE CALCULATIONS

Chemicals are used extensively in wastewater treatment plant operations. Wastewater treatment plant operators add chemicals to various unit processes for slime-growth control, corrosion control, odor control, grease removal, BOD reduction, pH control, sludge-bulking control, ammonia oxidation, and bacterial reduction, among other reasons. To apply any chemical dose correctly it is important to be able to make certain dosage calculations. Chemical dosages are measured in ppm (parts per million) or mg/L (milligrams per liter). Parts per million (ppm) is always a comparison of weight (pound per million pounds). One pound of chemical added to 1 million pounds of water would be a dosage of 1 ppm. Because each gallon of water weighs

8.34 lb, 1 million gallons of water weighs 8.34 million lb and would require 8.34 lb of chemical to obtain a dosage of 1 ppm. Milligrams per liter (mg/L) is the metric term for a dosage equal to ppm.

$$1 \text{ gallon} = 8.34 \text{ lb}$$

$$1 \text{ ppm} = 1 \text{ mg/L}$$

The number of pounds of chemical needed to achieve a certain dosage can be determined by multiplying the ppm by the number of millions of gallons treated and then by 8.34 lb/gal. The amount of water to be treated must always be in terms of millions of gallons (MGD).

$$\text{lb/day} = \text{mg/L} \times \text{MGD} \times 8.34 \text{ lb/gal}$$

One of the most frequently used calculations in wastewater mathematics is the conversion of milligrams per liter (mg/L) concentration to pounds per day (lb/day) or pounds (lb) dosage or loading. The general types of mg/L to lb/day or lb calculations are for chemical dosage, BOD, chemical oxygen demand (COD), suspended solids (SS) loading/removal, pounds of solids under aeration, and waste activated sludge (WAS) pumping rate. These calculations are usually made using either of the following equations:

$$\text{Dosage (lb/day)} = \text{Concentration (mg/L)} \times \text{Flow (MGD)} \times 8.34 \text{ lb/gal} \quad (16.18)$$

$$\text{Dosage (lb)} = \text{Concentration (mg/L)} \times \text{Volume (MG)} \times 8.34 \text{ lb/gal} \quad (16.19)$$

Note: If the mg/L concentration represents concentration in a flow, then million gallons per day (MGD) flow is used as the second factor; however, if the concentration pertains to a tank or pipeline volume, then million gallons (MG) volume is used as the second factor.

Dosage Formula Pie Chart

When converting pounds (lb) or mg/L, million gallons (MG) and 8.34 are key parameters. The pie chart shown in Figure 16.4 and the steps listed below can be helpful in finding lb or mg/L.

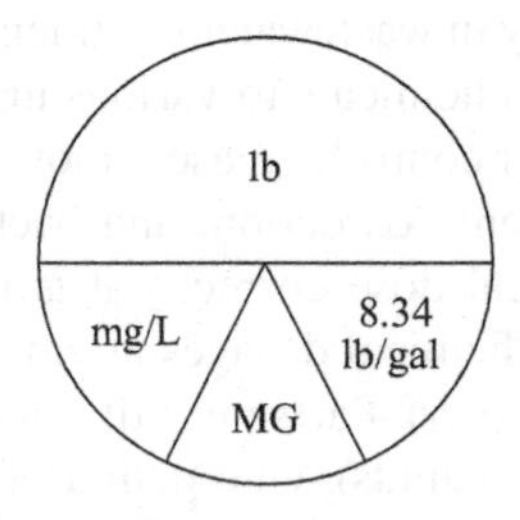

FIGURE 16.4 Dosage formula pie chart.

1. Determine what unit the question is asking you to find (lb or mg/L).
2. Physically cover or hide the area of the chart containing the desired unit. Write the desired unit down alone on one side of the equal sign to begin the necessary equation (e.g., lb =).
3. Look at the remaining uncovered areas of the circle. These *exactly* represent the other side of the equals sign in the necessary equation. If the unit above the center line is not covered, your equation will have a top (or numerator) and a bottom (or denominator), just like the pie chart. Everything above the center line goes in the numerator (the top of the equation) and everything below the center lines goes in the denominator (the bottom of the equation). Remember that all units below the line are *always multiplied together*; for example, if you are asked to find the dosage in mg/L, you would cover mg/L in the pie chart and write it down on one side of the equal sign to start your equation, like this:

$$\text{mg/L} =$$

The remaining portions of the pie chart are lb on top divided by MGD × 8.34 lb/gal on the bottom and would be written like this:

$$\text{mg/L} = \frac{\text{lb}}{\text{MGD} \times 8.34 \text{ lb/gal}}$$

If the area above the center line is covered, the right side of your equation will be made up of only the units below the center line. Remember that all units below the line are always multiplied together.

If you are asked, for example, to find the number of pounds needed, you would cover lb in the pie chart and write it down on one side of the equal sign to start your equation, like this:

$$\text{lb} =$$

All of the remaining areas of the pie chart are together on one line (below the center line of the circle), multiplied together on the other side of the equal sign, and written like this:

$$\text{lb} = \text{mg/L} \times \text{MGD} \times 8.34 \text{ lb/gal}$$

Chlorine Dosage

Chlorine is a powerful oxidizer commonly used in water treatment for purification and in wastewater treatment for disinfection, odor control, bulking control, and other applications. When chlorine is added to a unit process, we want to ensure that a measured amount is added. The amount of chemical added or required can be specified in two ways:

- Milligrams per liter (mg/L)
- Pounds per day (lb/day)

To convert from mg/L (or ppm) concentration to lb/day, we use Equation 16.20:

$$\text{Chlorine (lb/day)} = \text{Chlorine (mg/L)} \times \text{Flow (MGD)} \times 8.34 \text{ lb/gal} \quad (16.20)$$

Note: At one time, it was normal practice to use the expression *parts per million* (ppm) as an expression of concentration, as 1 mg/L = 1 ppm; however, current practice is to use mg/L as the preferred expression of concentration.

■ Example 16.21

Problem: Determine the chlorinator setting (lb/day) required to treat a flow of 8 MGD with a chlorine dose of 6 mg/L.

Solution:

$$\text{Chlorine (lb/day)} = \text{Chlorine (mg/L)} \times \text{Flow (MGD)} \times 8.34 \text{ lb/gal}$$

$$\text{Chlorine (lb/day)} = 6 \text{ mg/L} \times 8 \text{ MGD} \times 8.34 \text{ lb/gal} = 400 \text{ lb/day}$$

■ Example 16.22

Problem: What should the chlorinator setting be (in lb/day) to treat a flow of 3 MGD if the chlorine demand is 12 mg/L and a chlorine residual of 2 mg/L is desired?

Note: The chlorine demand is the amount of chlorine used in reacting with various components of the wastewater such as harmful organisms and other organic and inorganic substances. When the chlorine demand has been satisfied, these reactions stop.

Solution: Remember that

$$\text{Chlorine (lb/day)} = \text{Chlorine (mg/L)} \times \text{Flow (MGD)} \times 8.34 \text{ lb/gal}$$

To find the unknown value of lb/day, we must first determine chlorine dose. To do this we must use Equation 16.21:

$$\text{Chlorine dose (mg/L)} = \text{Chlorine demand (mg/L)} + \text{Chlorine residual (mg/L)} \quad (16.21)$$

$$\text{Chlorine dose} = 12 \text{ mg/L} + 2 \text{ mg/L} = 14 \text{ mg/L}$$

Then we can make the mg/L to lb/day calculation:

$$\text{Chlorine (lb/day)} = 12 \text{ mg/L} \times 3 \text{ MGD} \times 8.34 \text{ lb/gal} = 300 \text{ lb/day}$$

■ Example 16.23

Problem: How many pounds per day of chlorine are required to provide a dosage of 2.3 mg/L in 820,000 gpd?

Solution: Change gpd to MGD:

$$820{,}000 \text{ gpd} = 0.82 \text{ MGD}$$

Calculate the pounds per day:

$$\text{Chlorine (lb/day)} = 2.3 \text{ mg/L} \times 0.82 \times 8.34 = 15.7 \text{ lb/day}$$

Hypochlorite Dosage

At many wastewater facilities, sodium hypochlorite or calcium hypochlorite is used instead of chlorine. The reasons for substituting hypochlorite for chlorine vary; however, due to the passage of stricter hazardous chemicals regulations by the Occupational Safety and Health Administration (OSHA) and the U.S. Environmental Protection Agency (USEPA), many facilities are deciding to substitute nonhazardous hypochlorite for the hazardous chemical chlorine. Obviously, the potential liability involved with using deadly chlorine is also a factor involved in the decision to substitute it with a less toxic chemical substance.

For whatever reason, when a wastewater treatment plant decides to substitute chlorine for hypochlorite, the wastewater operator needs to be aware of the differences between the two chemicals. Chlorine is a hazardous material. Chlorine gas is used in wastewater treatment applications at 100% available chlorine. This is an important consideration to keep in mind when making or setting chlorine feed rates. For example, if the chlorine demand and residual require 100 lb/day chlorine, the chlorinator setting would be just that—100 lb/24 hr. Hypochlorite is less hazardous than chlorine; it is similar to strong bleach and comes in two forms: dry calcium hypochlorite (often referred to as HTH) and liquid sodium hypochlorite. Calcium hypochlorite contains about 65% available chlorine; sodium hypochlorite contains about 12 to 15% available chlorine (in industrial strengths).

> ***Note:*** Because neither type of hypochlorite is 100% pure chlorine, more lb/day must be fed into the system to obtain the same amount of chlorine for disinfection. This is an important economical consideration for those facilities thinking about substituting hypochlorite for chlorine. Some studies indicate that such a switch can increase overall operating costs by up to 3 times the cost of using chlorine.

To calculate the lb/day hypochlorite required, a two-step calculation is necessary:

$$\text{Chlorine (lb/day)} = \text{Chlorine (mg/L)} \times \text{Flow (MGD)} \times 8.34 \text{ lb/gal} \tag{6.22}$$

$$\text{Hypochlorite (lb/day)} = \frac{\text{Chlorine (lb/day)}}{\text{\% Available}} \times 100 \tag{6.23}$$

■ Example 16.24

Problem: A total chlorine dosage of 10 mg/L is required to treat a particular wastewater. If the flow is 1.4 MGD and the hypochlorite has 65% available chlorine, how many lb/day of hypochlorite will be required?

Solution: First calculate the lb/day chlorine required using the mg/L to lb/day equation:

$$\text{Chlorine (lb/day)} = \text{Chlorine (mg/L)} \times \text{Flow (MGD)} \times 8.34\ \text{lb/gal}$$

$$\text{Chlorine (lb/day)} = 10\ \text{mg/L} \times 1.4\ \text{MGD} \times 8.34\ \text{lb/gal} = 117\ \text{lb/day}$$

Then calculate the lb/day hypochlorite required. Because only 65% of the hypochlorite is chlorine, more than 117 lb/day will be required:

$$\frac{117\ \text{lb/day}}{65\%\ \text{Available chlorine}} \times 100 = 180\ \text{lb/day hypochlorite}$$

■ Example 16.25

Problem: A tank is 40 ft in diameter and 20 ft high and is dosed with 50 ppm of chlorine. How many pounds of 70% HTH are needed?

Solution: Find the volume of the tank in cubic feet:

$$3.14 \times 20\ \text{ft} \times 20\ \text{ft} \times 20\ \text{ft} = 25{,}133\ \text{ft}^3$$

Change cubic feet to gallons:

$$25{,}133\ \text{ft}^3 \times 7.48\ \text{gal/ ft}^3 = 187{,}993\ \text{gal}$$

Change gallons to million gallons per day:

$$187{,}993\ \text{gal} = 0.188\ \text{MGD}$$

Find the pounds of chlorine:

$$50\ \text{ppm} \times 0.188\ \text{mg} \times 8.34\ \text{lb/gal} = 78.4\ \text{lb}$$

Change percent available to a decimal equivalent:

$$70\% = 0.70$$

Find pounds of HTH:

$$78.4\ \text{lb} \div 0.70 = 112\ \text{lb}$$

■ Example 16.26

Problem: A wastewater flow of 840,000 gpd requires a chlorine dose of 20 mg/L. If sodium hypochlorite (15% available chlorine) is to be used, how many pounds per day of sodium hypochlorite are required? How many gallons per day of sodium hypochlorite is this?

Solution: Calculate the pounds per day chlorine required:

$$\text{Chlorine (lb/day)} = \text{Chlorine (mg/L)} \times \text{Flow (MGD)} \times 8.34 \text{ lb/gal}$$

$$\text{Chlorine (lb/day)} = 20 \text{ mg/L} \times 0.84 \text{ MGD} \times 8.34 \text{ lb/gal} = 140 \text{ lb/day}$$

Calculate the pounds per day sodium hypochlorite:

$$\frac{140 \text{ lb/day chlorine}}{15\% \text{ available chlorine}} \times 100 = 933 \text{ lb/day}$$

Calculate the gallons per day sodium hypochlorite:

$$933 \text{ lb/day} \div 8.34 \text{ lb/gal} = 112 \text{ gal/day}$$

■ Example 16.27

Problem: How many pounds of chlorine gas are necessary to treat 5,000,000 gal of wastewater at a dosage of 2 mg/L?

Solution:

$$(5 \times 10^6 \text{ gal}) \times 2 \text{ mg/L} \times 8.34 \text{ lb/gal} = 83 \text{ lb chlorine}$$

■ Example 16.28

Problem: A chlorine pump is feeding 10% bleach at a dosage of 5 mg/L. If 2,000,000 gal are treated in 18 hr, how many gallons per hour is the pump feeding?

Solution: Change gallons to million gallons:

$$2{,}000{,}000 \text{ gal} = 2.0 \text{ MG}$$

Find the pounds of chlorine:

$$5 \text{ mg/L} \times 2.0 \text{ MG} \times 8.34 \text{ lb/gal} = 83.4 \text{ lb}$$

Change percent available to a decimal equivalent:

$$10\% = 0.10$$

Find pounds of bleach:

$$83.4 \text{ lb} \div 0.10 = 834 \text{ lb}$$

Find gallons of bleach:

$$834 \text{ lb} \div 8.34 \text{ lb/gal} = 100 \text{ gal}$$

Find gallons per hour:

$$100 \text{ gal} \div 18 \text{ hr} = 5.5 \text{ gal/hr}$$

■ Example 16.29

Problem: A 12-in. pipe is 1875 ft long and must be disinfected with 60 ppm of 65% HTH. How many pounds of HTH are needed?

Solution: Find the volume of the pipe in gallons:

$$\text{Volume} = D^2 \times 0.0408 \times \text{Length} = (12 \text{ in.})^2 \times 0.0408 \times 1875 \text{ ft} = 11{,}016 \text{ gal}$$

Change gallons to MGD:

$$11{,}016 \text{ gal} \div 1{,}000{,}000 = 0.011 \text{ MGD}$$

Find pounds of chlorine:

$$\text{Chlorine (lb)} = 60 \text{ ppm} \times 0.011 \text{ MGD} \times 8.34 \text{ lb/gal} = 5.5 \text{ lb}$$

Change percent available to a decimal equivalent:

$$65\% = 0.65$$

Find pounds of HTH:

$$5.5 \text{ lb chlorine} \div 0.65 = 8.5 \text{ lb HTH}$$

Liquid chemical dosages can be calculated to determine the gallons per day. Chemical feed pumps are calibrated using milliliters per minute (mL/min). If you take 3650 mL/gal and divide it by 1440 min/day, the conversion for gal/day can be determined.

$$3650 \text{ mL/gal} \div 1440 \text{ min/day} = 2.5 \text{ mL/min/gal/day}$$

■ Example 16.30

Problem: A 20% available fluoride solution is used to dose 2,000,000 gpd at 440 ppb (parts per billion). How many milliliters per minute is the pump feeding?

Solution: Change 440 ppb to ppm:

$$440 \text{ ppb} \div 1000 = 0.44 \text{ ppm (mg/L)}$$

Change 2,000,000 gpd to MGD:

$$2{,}000{,}000 \text{ gpd} \div 1{,}000{,}000 = 2 \text{ MGD}$$

Find pounds of fluoride:

$$\text{Fluoride} = 0.44 \text{ ppm} \times 2.0 \text{ MGD} \times 8.34 \text{ lb/gal} = 7.3 \text{ lb/day}$$

Change percent available to a decimal equivalent:

$$20\% = 0.2$$

Find pounds per day of fluoride solution:

$$7.3 \text{ lb/day} \div 0.2 = 36.5 \text{ lb/day}$$

Find gallons per day of fluoride:

$$36.5 \text{ lb/day solution} \div 8.34 \text{ lb/gal} = 4.4 \text{ gpd}$$

Change gallons per day to milliliters per minute:

$$4.4 \text{ gpd} \times 2.6 = 11.4 \text{ mL/min}$$

■ Example 6.31

Problem: An 18% available alum solution is used to dose 800,000 gpd at 22 mg/L. How many milliliters per minute is the pump feeding?

Solution: Change 800,000 gpd to MGD:

$$800{,}000 \text{ gpd} = 0.8 \text{ MGD}$$

Find pounds of alum:

$$\text{Alum} = 22 \text{ mg/L} \times 0.8 \text{ MGD} \times 8.34 \text{ lb/gal} = 147 \text{ lb/day}$$

Change percent available to a decimal equivalent:

$$18\% = 0.18$$

Find pounds per day of alum solution:

$$147 \text{ lb/day} \div 0.18 = 817 \text{ lb/day}$$

Find gallons per day of alum:

$$817 \text{ lb/day solution} \div 8.34 \text{ lb/gal} = 98 \text{ gpd}$$

Change gallons per day to milliliters per minute:

$$98 \text{ gpd} \times 2.6 = 255 \text{ mL/min}$$

LABORATORY EXAMPLES

■ Example 16.32

Problem: A laboratory solution is made using 50 mg of sodium chloride (NaCl) dissolved in a 1-L volumetric flask filled to the mark. What is the milligram per liter concentration of the solution?

Solution:

$$50 \text{ mg} \div 1 \text{ L} = 50 \text{ mg/L}$$

■ Example 16.33

Problem: If 35 lb of a chemical are added to 146 lb of water, what is the percent strength by weight?

Solution:

$$\% \text{ Strength} = \frac{\text{Weight of chemical}}{\text{Weight of water} + \text{Weight of chemical}} \times 100$$

$$= \frac{35 \text{ lb}}{146 \text{ lb} + 35 \text{ lb}} \times 100 = 19.34\%$$

■ Example 16.34

Problem: If 1 L of 0.1-N HCL is required but 12-N HCL is on hand, how many milliliters of the 12-N HCL are required to make the 1 L?

Solution:

$$N_1V_1 = N_2V_2$$

$$0.1N \times 1000 \text{ mL} = 12N \times V_2$$

$$\frac{0.1N \times 1000 \text{ mL}}{12N} = V_2$$

$$8.33 \text{ mL} = V_2$$

■ **Example 16.35**

Problem: If 260 mL of 3-*N* NaOH are diluted to 1000 mL, what is the new normality of the solution?

Solution:

$$N_1V_1 = N_2V_2$$

$$3\,N \times 260\text{ mL} = N_2 \times 1000\text{ mL}$$

$$\frac{3\,N \times 260\text{ mL}}{1000\text{ mL}} = N_2$$

$$0.78 = N_2$$

■ **Example 16.36**

Problem: If 600 mL of 10-*N* NaOH is diluted to 1 L, what is the new normality of the solution?

Solution:

$$N_1V_1 = N_2V_2$$

$$10\,N \times 600\text{ mL} = N_2 \times 1000\text{ mL}$$

$$\frac{10\,N \times 600\text{ mL}}{1000\text{ mL}} = N_2$$

$$6 = N_2$$

■ **Example 16.37**

Problem: Given 22 mL of 30-*N* HCl, how many milliliters of water should be added to make 1.2-*N* HCL?

Solution:

$$N_1V_1 = N_2V_2$$

$$30\,N \times 22\text{ mL} = 12\,N \times V_2$$

$$\frac{30\,N \times 22\text{ mL}}{12\,N} = V_2$$

$$550 = V_2$$

■ **Example 16.38**

Problem: A 0.1-*N* solution is needed to conduct an analysis, but a 1.2-*N* solution is on hand. How many milliliters of the 1.2-*N* solution are required to make 1 L of a 0.1-*N* solution?

Solution:

$$N_1V_1 = N_2V_2$$

$$0.1N \times 1000 \text{ mL} = 1.2N \times V_2$$

$$\frac{0.1N \times 1000 \text{ mL}}{1.2N} = V_2$$

$$83.3 \text{ mL} = V_2$$

■ Example 16.39

Problem: A 0.1-*N* solution is needed to conduct an analysis, but a 2.1-*N* solution is on hand. How many milliliters of the 2.1-*N* solution are required to make 1 L of a 0.1-*N* solution?

Solution:

$$N_1V_1 = N_2V_2$$

$$0.1N \times 1000 \text{ mL} = 2.1N \times V_2$$

$$\frac{0.1N \times 1000 \text{ mL}}{2.1N} = V_2$$

$$47.6 \text{ mL} = V_2$$

■ Example 16.40

Problem: If 475 mL of 5-*N* NaOH are diluted to 1 L, what is the new normality of the solution?

Solution:

$$N_1V_1 = N_2V_2$$

$$5N \times 450 \text{ mL} = N_2 \times 1000 \text{ mL}$$

$$\frac{5N \times 450 \text{ mL}}{1000 \text{ mL}} = N_2$$

$$2.375 = N_2$$

■ Example 16.41

Problem: How many milliliters of water should be added to 10 mL of 15-*N* H_2SO_4 to make 0.4-*N* H_2SO_4?

Solution:

$$N_1V_1 = N_2V_2$$

$$15\,N \times 10\text{ mL} = 0.4\,N \times V_2$$

$$\frac{15\,N \times 10\text{ mL}}{0.4\,N} = V_2$$

$$375\text{ mL} = V_2$$

■ EXAMPLE 16.42

Problem: A 0.2-*N* solution is needed but a 2.7-*N* solution is on hand. How many milliliters of the 2.7-*N* solution are needed to make 0.5 L of 0.2-*N* solution?

Solution:

$$N_1V_1 = N_2V_2$$

$$2.7\,N \times 500\text{ mL} = 0.2\,N \times V_2$$

$$\frac{2.7\,N \times 500\text{ mL}}{0.2\,N} = V_2$$

$$6750\text{ mL} = V_2$$

17 Basic Mathemathics Practice Problems

BASIC MATH OPERATIONS

Decimal Operations

1. $90.5 \times 7.3 =$

2. $9.556 \times 1.03 =$

3. $13 \div 14.3 =$

4. $8.2 \div 0.96 =$

5. $2 \div 0.053 =$

6. Convert 3/4 into a decimal.

7. Convert 1/6 into a decimal.

8. Convert 3/8 into a decimal.

9. Convert 0.13 into a fraction.

10. Convert 0.9 into a fraction.

11. Convert 0.75 into a fraction.

12. Convert 0.245 into a fraction.

Percentage Calculations

13. Convert 15/100 into a percent.

14. Convert 122/100 into a percent.

15. Convert 1.66 into a percent.

16. Convert 4/7 into a percent.

17. A 100% decline from 66 leaves us with how much?

Solving for an Unknown

18. If $x - 6 = 2$, what is the value of x?

19. If $x - 4 = 9$, what is the value of x?

20. If $x - 8 = 17$, what is the value of x?

21. If $x + 10 = 15$, what is the value of x?

22. Find x when $x/3 = 2$

23. Solve for x when $x/4 = 10$

24. If $4x = 8$, what is the value of x?

25. If $6x = 15$, what is the value of x?

26. If $x + 10 = 2$, what is the value of x?

27. Find x if $x - 2 = -5$

28. If $x + 4 = -8$, what is the value of x?

29. If $x - 10 = -14$, find x.

30. If $0.5x - 1 = -6$, find x.

31. If $9x + 1 = 0$, what is the value of x?

32. How much is x^2 if $x = 6$?

33. If $x = 3$, what is the value of x^4?

34. If $x = 10$, what is the value of x^0?

Ratio and Proportion

35. If an employee was out sick on 6 of 96 workdays, what is that employee's ratio of sick days to days worked?

36. Find x when $2{:}x = 5{:}15$

Area of Rectangles

37. What is the area of a rectangle 3 ft long by 3 ft wide?

38. What is the area of a rectangle 5 in. long by 3 in. wide?

39. What is the area of a rectangle 5 yd long by 1 yd wide?

Circumference and Area of Circles

40. Find the circumference of a circle whose diameter is 14 ft.

41. If the circumference of a circle is 8 in., what is the diameter?

42. If the radius of a circle is 7 in., what is its area?

43. If the diameter of a circle is 10 in., what is its area?

FUNDAMENTAL WATER/WASTEWATER OPERATIONS

Tank Volume Calculations

1. The diameter of a tank is 70 ft. If the water depth is 25 ft, what is the volume of water in the tank in gallons?

2. A tank is 60 ft in length, 20 ft wide, and 10 ft deep. Calculate the volume of the tank in cubic feet.

3. A tank 20 ft wide by 60 ft long is filled with water to a depth of 12 ft. What is the volume of the water in the tank in gallons?

4. What is the volume of water in a tank, in gallons, if the tank is 20 ft wide by 40 ft long and contains water to a depth of 12 ft?

5. A tank has a diameter of 60 ft and a depth of 12 ft. Calculate the volume of water in the tank in gallons.

6. What is the volume of water in a tank, in gallons, if the tank is 20 ft wide by 50 ft long and contains water to a depth of 16 ft?

Channel and Pipeline Capacity Calculations

7. A rectangular channel is 340 ft in length, 4 ft in depth, and 6 ft wide. What is the volume of water in cubic feet?

8. A replacement section of 10-in. pipe is to be sandblasted before it is put into service. If the length of pipeline is 1600 ft, how many gallons of water will be required to fill the pipeline?

9. A trapezoidal channel is 800 ft in length, 10 ft wide at the top, and 5 ft wide at the bottom, with a distance of 4 ft from the top edge to the bottom along the sides. Calculate the gallon volume.

10. A section of 8-in.-diameter pipeline is to be filled with treated water for distribution. If the pipeline is 2250 ft in length, how many gallons of water will be distributed?

11. A channel is 1200 ft in length by 5 ft wide and carries water 4 ft in depth. What is the volume of water in gallons?

Miscellaneous Volume Calculations

12. A pipe trench is to be excavated that is 4 ft wide, 4 ft deep, and 1200 ft long. What is the volume of the trench in cubic yards?

13. A trench is to be excavated that is 3 ft wide, 4 ft deep, and 500 yd long. What is the cubic yard volume of the trench?

14. A trench is 300 yd long, 3 ft wide, and 3 ft deep. What is the volume of the trench in cubic feet?

15. A rectangular trench is 700 ft long, 6.5 ft wide, and 3.5 ft deep. What is the volume of the trench in cubic feet?

16. The diameter of a tank is 90 ft. If the water depth in the tank is 25 ft, what is the volume of water in the tank in gallons?

17. A tank is 80 ft long, 20 ft wide, and 16 ft deep. What is the volume of the tank in cubic feet?

18. How many gallons of water will it take to fill an 8-in.-diameter pipe that is 4000 ft in length?

19. A trench is 400 yd long, 3 ft wide, and 3 ft deep. What is the volume of the trench in cubic feet?

20. A trench is to be excavated. If the trench is 3 ft wide, 4 ft deep, and 1200 ft long, what is the volume of the trench in cubic yards?

21. A tank is 30 ft wide and 80 ft long. If the tank contains water to a depth of 12 ft, how many gallons of water are in the tank?

22. What is the volume of water, in gallons, contained in a 3000-ft section of channel if the channel is 8 ft wide and the water depth is 3.5 ft?

23. A tank has a diameter of 70 ft and a depth of 19 ft. What is the volume of water in the tank in gallons?

24. If a tank is 25 ft in diameter and 30 ft deep, how many gallons of water will it hold?

Flow, Velocity, and Conversion Calculations

25. A channel 44 in. wide has water flowing to a depth of 2.4 ft. If the velocity of the water is 2.5 fps, what is the flow in the channel in cubic feet per minute?

26. A tank is 20 ft long and 12 ft wide. With the discharge valve closed, the influent to the tank causes the water level to rise 0.8 feet in 1 minute. What is the flow to the tank in gallons per minute?

27. A trapezoidal channel is 4 ft wide at the bottom and 6 ft wide at the water surface. The water depth is 40 in. If the flow velocity through the channel is 130 ft/min, what is the flow rate through the channel in cubic feet per minute?

28. An 8-in.-diameter pipeline has water flowing at a velocity of 2.4 fps. What is the flow rate through the pipeline in gallons per minute? Assume the pipe is flowing full.

29. A pump discharges into a 3-ft-diameter container. If the water level in the container rises 28 inches in 30 seconds, what is the flow into the container in gallons per minute?

30. A 10-in.-diameter pipeline has water flowing at a velocity of 3.1 fps. What is the flow rate through the pipeline, in gallons per minute, if the water is flowing at a depth of 5 in.?

31. A channel has a rectangular cross-section. The channel is 6 ft wide with water flowing to a depth of 2.6 ft. If the flow rate through the channel is 14,200 gpm, what is the velocity of the water in the channel in feet per second?

32. An 8-in.-diameter pipe flowing full delivers 584 gpm. What is the velocity of flow in the pipeline in feet per second?

33. A special dye is used to estimate the velocity of flow in an interceptor line. The dye is injected into the water at one pumping station and the travel time to the first manhole 550 ft away is noted. The dye first appears at the downstream manhole in 195 seconds. The dye continues to be visible until the total elapsed time is 221 seconds. What is the velocity of flow through the pipeline in feet per second?

34. The velocity in a 10-in.-diameter pipeline is 2.4 fps. If the 10-in. pipeline flows into an 18-in.-diameter pipeline, what is the velocity in the 8-in. pipeline in feet per second?

35. A float travels 500 ft in a channel in 1 min, 32 sec. What is the estimated velocity in the channel in feet per second?

36. The velocity in an 8-in.-diameter pipe is 3.2 fps. If the flow then travels through a 10-in.-diameter section of pipeline, what is the velocity in the 10-in. pipeline in feet per second?

Average Flow Rates

37. The flows listed below were recorded for a week. What was the average daily flow rate for the week?

    Monday—4.8 MGD
    Tuesday—5.1 MGD
    Wednesday—5.2 MGD
    Thursday—5.4 MGD
    Friday—4.8 MGD
    Saturday—5.2 MGD
    Sunday—4.8 MGD

38. The totalizer reading for the month of September was 121.4 MG. What was the average daily flow for the month of September?

Flow Conversions

39. Convert 0.165 MGD to gpm.

40. The total flow for one day at a plant was 3,335,000 gal. What was the average flow for that day in gallons per minute?

41. Express a flow of 8 cfs in terms of gpm.

42. What is 35 gps expressed as gpd?

43. Convert a flow of 4,570,000 gpd to cfm.

44. What is 6.6 MGD expressed as cfs?

45 Express 445,875 cfd as gpm.

46. Convert 2450 gpm to gpd.

General Flow and Velocity Calculations

47. A channel has a rectangular cross-section. The channel is 6 ft wide with water flowing to a depth of 2.5 ft. If the flow rate through the channel is 14,800 gpm, what is the velocity of the water in the channel in feet per second?

48. A channel 55 in. wide has water flowing to a depth of 3.4 ft. If the velocity of the water is 3.6 fps, what is the flow in the channel in cubic feet per minute?

49. The following flows were recorded over a period of 3 months: June, 102.4 MG; July, 126.8 MG; and August, 144.4 MG. What was the average daily flow for this 3-month period?

50. A tank is 12 ft by 12 ft. With the discharge valve closed, the influent to the tank causes the water level to rise 8 inches in 1 minute. What is the flow to the tank in gallons per minute?

51. An 8-in.-diameter pipe flowing full delivers 510 gpm. What is the velocity of flow in the pipeline in feet per second?

52. Express a flow of 10 cfs in terms of gallons per minute.

53. The totalizer reading for the month of December was 134.6 MG. What was the average daily flow for the month of September?

54. What is 5.2 MGD expressed as cubic feet per second?

55. A pump discharges into a 3-ft-diameter container. If the water level in the container rises 20 inches in 30 seconds, what is the gpm flow into the container?

56. Convert a flow of 1,825,000 gpd to cubic feet per minute.

57. A 6-in.-diameter pipeline has water flowing at a velocity of 2.9 fps. What is the flow rate through the pipeline in gallons per minute?

58. The velocity in a 10-in. pipeline is 2.6 fps. If the 10-in. pipeline flows into an 8-in.-diameter pipeline, what is the velocity in the 8-in. pipeline in feet per second?

59. Convert 2225 gpm to gpd.

60. The total flow for one day at a plant was 5,350,000 gal. What was the average flow for that day in gallons per minute?

Chemical Dosage Calculations

61. Determine the chlorinator setting, in pounds per day, necessary to treat a flow of 5.5 MGD with a chlorine dose of 2.5 mg/L.

62. To dechlorinate a wastewater, sulfur dioxide is to be applied at a level 4 mg/L more than the chlorine residual. What should be the sulfonator feed rate, in pounds per day, for a flow of 4.2 MGD with a chlorine residual of 3.1 mg/L?

63. What should be the chlorinator setting, in pounds per day, to treat a flow of 4.8 MGD if the chlorine demand is 8.8 mg/L and a chlorine residual of 3 mg/L is desired?

64. A total chlorine dosage of 10 mg/L is required to treat the water in a unit process. If the flow is 1.8 MGD and the hypochlorite has 65% available chlorine, how many pounds per day of hypochlorite will be required?

65. The chlorine dosage at a plant is 5.2 mg/L. If the flow rate is 6,250,000 gpd, what is the chlorine feed rate in pounds per day?

66. A storage tank is to be disinfected with 60 mg/L of chlorine. If the tank holds 86,000 gal, how many pounds of chlorine gas will be needed?

67. To neutralize a sour digester, 1 lb of lime is to be added for every pound of volatile acids in the digester liquor. If the digester contains 225,000 gal of sludge with a volatile acid level of 2,220 mg/L, how many pounds of lime should be added?

68. A flow of 0.83 MGD requires a chlorine dosage of 8 mg/L. If the hypochlorite has 65% available chlorine, how many pounds per day of hypochlorite will be required?

Appendix: Solutions to Chapter 17 Problems

BASIC MATH OPERATIONS

1. 660.65
2. 9.84268
3. 0.91
4. 8.5
5. 37.7
6. 0.75
7. 0.167
8. 0.375
9. 13/100
10. 9/10
11. 3/4
12. 49/200
13. 15%
14. 122%
15. 166%
16. 57.1%
17. 0
18. $x - 6 = 2$
    $x - 6 + 6 = 2 + 6$
    $x = 8$
19. $x - 4 = 9$
    $x = 13$
20. $x = 17 + 8 = 25$
21. $x = 15 - 10 = 5$
22. $3(x/3) = 2 \times x$
    $x = 6$
23. $4(x/4) = 4 \times 10$
    $x = 40$
24. $x - 8/4 - 2$
25. $x = 15/6 = 2.5$
26. $x + 10 - 10 = 2 - 10$
    $x = -8$
27. $x - 2 + 2 = -5 + 2$
    $x = -3$
28. $x + 4 - 4 = -8 - 4$
    $x = -12$

29. $x - 10 + 10 = -14 + 10$
    $x = -4$
30. $0.5x - 1 + 1 = -6 + 1$
    $x = -10$
31. $9x + 1 - 1 = 0 - 1$
    $x = -1/9$
32. $x^2 = 6^2 = 6 \times 6 = 36$
33. $3 \times 3 \times 3 \times 3 = 81$
34. 1
35. 36:1
36. $2(15) = 5x$
    $30 = 5x$
    $6 = x$
37. 3 ft × 3 ft = 9 ft^2
38. 5 in. × 3 in. = 15 in.^{2}
39. 5 yd × 1 yd = 5 yd^2
40. Circumference = $\pi \times D$ = 3.14 × 14 ft = 44 ft
41. 8 in. = $\pi \times D$
    8 in. = 3.14 × D
    8 in./3.14 = D
    2.5 in. = D
42. Area = $\pi \times r^2$ = 3.14 × (6 in.)^2 = 3.14 × 36 in.^{2} = 113 in.^{2}
43. Area = $\pi \times r^2$ = 3.14 × (5 in.)^2 = 3.14 × 25 in.^{2} = 78.5 in.^{2}

FUNDAMENTAL WATER/WASTEWATER OPERATIONS

1. 0.785 × (70 ft)^2 × 25 ft × 7.48 gal/ft^3 = 719,295.5 gal
2. 60 ft × 20 ft × 10 ft = 12,000 ft^3
3. 20 ft × 60 ft × 12 ft × 7.48 gal/ft^3 = 107,712 gal
4. 20 ft × 40 ft × 12 ft × 7.48 gal/ft^3 = 71,808 gal
5. 0.785 × (60 ft)^2 × 12 ft × 7.48 gal/ft^3 = 253,662 gal
6. 20 ft × 50 ft × 16 ft × 7.48 gal/ft^3 = 119,680 gal
7. 4 ft × 6 ft × 340 ft = 8160 ft^3
8. 0.785 × (0.83 ft)^2 × 1600 ft × 7.48 gal/ft^3 = 6472 gal
9. [(5 ft + 10 ft)/2)] × 4 ft × 800 ft × 7.48 gal/ft^3 = 179,520 gal
10. 0.785 × (0.66)^2 × 2250 ft × 7.48 gal/ft^3 = 5755 gal
11. 5 ft × 4 ft × 1200 ft × 7.48 gal/ft^3 = 179,520 gal
12. $$\frac{4\text{ ft} \times 4\text{ ft} \times 1200\text{ ft}}{27\text{ ft}^3/\text{yd}^3} = 711\text{ yd}^3$$
13. 500 yd × 1 yd × 1.33 yd = 665 yd^3
14. 900 ft × 3 ft × 3 ft = 8100 ft^3
15. 700 ft × 6.5 ft × 3.5 ft = 15,925 ft^3
16. 0.785 × (90 ft)^2 × 25 ft × 7.48 gal/ft^3 = 1,189,040 gal
17. 80 ft × 16 ft × 20 ft = 25,600 ft^3
18. 0.785 × (0.67 ft)^2 × 4000 ft × 7.48 gal/ft^3 = 10,543 gal

19. 1200 ft × 3 ft × 3 ft) = 10,800 ft^3
20. $\frac{3 \text{ ft} \times 4 \text{ ft} \times 1200 \text{ ft}}{27 \text{ ft}^3/\text{yd}^3} = 533 \text{ yd}^3$
21. 30 ft × 80 ft × 12 ft × 7.48 gal/ft^3 = 215,424 gal
22. 8 ft × 3.5 ft × 3000 ft × 7.48 gal/ft^3 = 628,320 gal
23. 0.785 × $(70 \text{ ft})^2$ × 19 ft × 7.48 gal/ft^3 = 546,665 gal
24. 0.785 × $(25 \text{ ft})^2$ × 30 ft × 7.48 gal/ft^3 = 110,096 gal
25. 2.4 ft × 3.7 ft × 2.5 fps × 60 sec/min = 1332 cfm
26. 20 ft × 12 ft × 0.8 fpm × 7.48 gal/ft^3 = 1436 gpm
27. [(4 ft + 6 ft)/2] × 3.3 ft × 130 fpm = 5 ft × 3.3 ft × 130 fpm = 2145 cfm
28. 0.785 × $(0.66 \text{ ft})^2$ × 2.4 fps × 7.48 gal/ft^3 × 60 sec/min = 368 gpm
29. 0.785 × $(3 \text{ ft})^2$ × 4.7 fpm × 7.48 gal/ft^3 = 248 gpm
30. 0.785 × $(0.83 \text{ ft})^2$ × 3.1 fps × 7.48 gal/ft^3 × 60 sec/min × 0.5 = 376 gpm
31. 6 ft × 2.6 ft × x fps × 60 sec/min × 7.48 gal/ft^3 = 14,200 gpm
    $x = 2.03$ ft
32. 0.785 × $(0.67 \text{ ft})^2$ × x fps × 60 sec/min × 7.48 gal/ft^3 = 584 gpm
    $x = 3.7$ fps
33. 550 ft ÷ 208 sec = 2.6 fps
34. 0.785 × $(0.83 \text{ ft})^2$ × 2.4 fps = 0.785 × $(0.67 \text{ ft})^2$ × x fps
    $x = 3.7$ fps
35. 500 ft/92 sec = 5.4 fps
36. 0.785 × $(0.67)^2$ × 3.2 fps = (0.785 × $(0.83 \text{ ft})^2$ × x fps
    $x = 2.1$ fps
37. 35.3 MGD/7 = 5 MGD
38. 121.4 MG/30 days = 4.0 MGD
39. 1,000,000 × 0.165 = 165,000 gpd
40. 3,335,000 gal ÷ 1440 min = 2316 gpm
41. 8 cfs × 7.48 gal/ft^3 × 60 sec/min = 3590 gpm
42. 35 gps × 60 sec/min × 1440 min/day = 3,024,000 gpd
43. $\frac{4{,}570{,}000}{1440 \text{ min/day} \times 7.48 \text{ gal/ft}^3} = 424 \text{ cfm}$
44. 6.6 MGD × 1.55 cfs/MGD = 10.2 cfs
45. $\frac{445{,}875 \text{ ft}^3/\text{day} \times 7.48 \text{ gal/ft}^3}{1440 \text{ min/day}} = 2316 \text{ gpm}$
46. 2450 gpm × 1440 min/day = 3,528,000 gpd
47. 6 ft × 2.5 ft × x fps × 60 sec/min × 7.48 gal/ft^3 = 14,800 gpm
    $x = 2.2$ fps
48. 4.6 ft × 3.4 ft × 3.6 fps × 60 sec/min = 3378 cfm
49. 373.6 MGD ÷ 92 days = 4.1 MGD
50. 12 ft × 12 ft × 0.67 fpm × 7.48 gal/ft^3 = 722 gpm
51. 0.785 × $(0.67 \text{ ft})^2$ × x fps × 60 sec/min × 7.48 gal/ft^3 = 510 gpm
    $x = 3.2$ fps
52. 10 cfs × 7.48 gal/ft^3 × 60 sec/min = 4488 gpm
53. 134.6 MG ÷ 31 days = 4.3 MGD

54. 5.2 MGD × 1.55 cfs/MGD = 8.1 cfs
55. 0.785 × (2 ft)^2 × 3.3 fpm × 7.48 gal/ft^3 = 77.5 gpm
56. $$\frac{1{,}825{,}000 \text{ gpd}}{1440 \text{ min/day} \times 7.48 \text{ gal/ft}^3} = 169 \text{ cfm}$$
57. 0.785 × (0.5 ft)^2 × 2.9 fps × 7.48 gal/ft^3 × 60 sec/min = 255 gpm
58. 0.785 × (0.83 ft)^2 × 2.6 fps = 0.785 × (0.67 ft)^2 × x fps
    x = 4.0 fps
59. 2225 gpm × 1440 min/day = 3,204,000 gpd
60. 5,350,000 gal ÷ 1440 min/day = 3715 gpm
61. 2.5 mg/L × 5.5 MGD × 8.34 lb/gal = 115 lb/day
62. 7.1 mg/L × 4.2 MGD × 8.34 lb/gal = 249 lb/day
63. 11.8 mg/L × 4.8 MGD × 8.34 lb/gal = 472 lb/day
64. $$\frac{10 \text{ mg/L} \times 1.8 \text{ MGD} \times 8.34 \text{ lb/gal}}{0.65} = 231 \text{ lb/day}$$
65. 41 mg/L × 6.25 MGD × 8.34 lb/gal = 214 lb/day
66. 60 mg/L × 0.086 MGD × 8.34 lb/gal = 43 lb
67. 2220 mg/L × 0.225 × 8.34 lb/gal = 4166 lb
68. $$\frac{8 \text{ mg/L} \times 0.83 \text{ MGD} \times 8.34 \text{ lb/gal}}{0.65} = 85 \text{ lb/day}$$

Index

L

M

N

O

P

Q

R

S

T

U

V

W

Printed in the United States
by Baker & Taylor Publisher Services

Printed in the United States
by Baker & Taylor Publisher Services